Heroes of Progress

Heroes of Progress

65 PEOPLE WHO CHANGED THE WORLD

By Alexander C. R. Hammond

Copyright © 2024 by the Cato Institute
All rights reserved.
Cato Institute is a registered trademark.

Print ISBN: 978-1-952223-67-9
eBook ISBN: 978-1-952223-68-6

Cover and interior design: Luis Ahumada Abrigo and
Guillermina Sutter Schneider
Cover illustrations: Alexandra Franzese
Interior illustrations: Yuriy Romanovich

Library of Congress Cataloging Number: 2023045722

Printed in the United States.

CATO INSTITUTE
1000 Massachusetts Ave. NW
Washington, DC 20001

www.cato.org

Dedicated to Linda Whetstone, a hero of liberty responsible for untold progress.

Contents

Foreword

Progress, be it economic, scientific, or cultural, depends on the human brain's ability to innovate and the institutions that make innovation and innovators possible and acceptable. Whether it's brilliant scientists whose groundbreaking discoveries saved billions of lives, inventors who devised technologies that spur humanity forward, or compassionate writers who awaken our empathy, history is full of individuals who have defied the odds and transformed countless lives for the better.

The valorization of people who conquer disease and poverty was once commonplace. In my childhood, the heroic biography of a pioneer whose efforts helped change the world for the better was a popular literary genre. Humanity's admiration for the progress made by scientists, inventors, or intellectuals also extended far beyond literature.

In 1847, a year after Richard Cobden's campaign to abolish the Corn Laws succeeded (an effort that led Britain down the path of free trade and greater economic prosperity), members of the public expressed their gratitude by raising over £80,000 (approximately $9.3 million in 2023 U.S. dollars) for him.

In France, about 3,400 streets, avenues, and boulevards and more than 360 higher education institutes are named after their countryman Louis Pasteur to honor his achievements in developing the germ theory of disease and creating the process of pasteurization.

Similarly, on April 12, 1955, when scientists announced that Jonas Salk's vaccine against polio was proven safe, the vaccine was celebrated as "more than a scientific achievement" and "an occasion for pride and jubilation," according to Salk's biographer Richard Carter. On that day, Carter notes, "people observed moments of silence, rang bells, honked horns, blew factory whistles, fired salutes . . . took the rest of the day off, closed schools or convoked fervid assemblies therein, drank toasts, hugged children, attended church, smiled at strangers, and forgave enemies."

Yet, despite these and other heroic stories, we appear to now live in a world where skepticism about humanity's ability to combat the most pressing problems of the day reigns supreme. Skepticism often breeds apathy and fatalism; consequently, many believe that the problems we face are too entrenched and complex to address effectively.

Ingratitude (a sin that, according to Dante, consigns perpetrators to the ninth circle of Hell) underpins much of our modern-day skepticism and has become ingrained in our media and universities. Whether it be a lack of historical knowledge or a post-1960s intellectual climate that seeks to downplay the achievements of our past, championing history's heroes has become rare, especially in mainstream culture.

Making an inspiring case for progress at this time of skepticism and historical ingratitude is no easy feat. Yet, by relentlessly outlining the extraordinary ability of individuals to shape our world for the better, Alexander Hammond does just that.

At its core, *Heroes of Progress* powerfully demonstrates that providing that people are free to speak, write, reason, and learn, progress is not only possible but likely.

The individuals profiled in this book stand on the shoulders of giants, often shared ideas within large teams, and relied on centuries, if not millennia, of aggregated knowledge. Nonetheless, whether it is the innovation or innovator, celebrating the groundbreaking discoveries and heroic feats that drastically improved the world for the better doesn't just make for good stories. Rather, these tales invoke a sense of gratitude and a can-do spirit in the reader that provide an antidote to fatalism and gloominess about the state of the world.

Similarly, *Heroes of Progress* reminds us that heroic tales of individuals changing the world for the better needn't be confined to the realm of fiction. Instead, these stories are all around us and are possible because free individuals are able to use their rationality and creativity for the betterment of humanity. Long may that continue.

Steven Pinker

Introduction

This book compiles the biographies of some of the most important people who have ever lived. It documents the key innovations, ideas, discoveries, or actions of 65 people whose work has shaped the world around us. When writing it, I hoped to create in the reader both a sense of gratitude for what has come before and inspiration for the possibilities of the future.

It is not a stretch to say that many of us might not be alive today if not for the work of the heroes profiled herein. And even if we were, without their contributions we would be far poorer, sicker, hungrier, less free, and much less knowledgeable of the world around us.

This book presents the biographies of the heroes of progress in chronological order based on their date of birth. This format provides the reader with a notion of how ideas, discoveries, actions, and inventions that have positively influenced humanity evolved over time. Furthermore, because the heroes profiled have contributed to a wide range of fields, there is no perfect way to sort them into different categories based on their contributions.

The chronological approach to documenting the heroes can also highlight that innovation and progress are not regular, linear, or guaranteed. Sometimes, nothing much happens for hundreds or even thousands of years. Rather, the recent phenomena of rapid and revolutionary innovation and the associated improvements in living standards globally are historical anomalies. Although a discussion of politics and economics is beyond the scope of this book, by presenting the heroes chronologically, it is hoped that the reader considers the types of societies and institutions that have been most conducive to bursts of innovation and progress.

This book aims to demonstrate that when given agency and freedom, human minds can accomplish extraordinary feats that can change the world for the better. With hard work, perseverance, freedom of agency, and luck, some of you may become "heroes of progress."

What Makes a "Hero of Progress"?

I have defined a hero of progress as someone who has saved or significantly improved the lives of millions of people. A hero could be a scientist who invented a vaccine that saved millions of lives, an agronomist who created a crop that fed billions of hungry people, or a thinker whose ideas brought about a richer, healthier, or more just society.

The process of innovation is usually incremental, collective, and part of a network that is built on decades, centuries, or millennia of previously acquired knowledge. For simplicity's sake, I have identified heroes as people whose work built on the accumulated wisdom of humankind to produce an invention, discovery, idea, or action that either directly or indirectly saved or improved the lives of millions. Doing so should not be seen as an attempt to disregard the achievements of those who came before. Indeed, in many cases, previous insights were crucial to the achievements of the heroes profiled in this volume.

Similarly, this book does not attempt to discuss every important innovation in history. Instead, it provides an overview of the people who played the clearest and most important roles in developing some of the most significant innovations that helped save or improve millions of lives. Innovations that built on the work of many predecessors but had no clear pioneer have been omitted, as profiling all the key people involved in their development would easily take up large sections of the book.

Moreover, the list of our heroes of progress is not meant to be all-encompassing. Likely, many more people, alive today or in the past, fit our narrow definition of a hero of progress but are not documented in this book. The exclusion of someone who could fit our definition of such a hero can be attributed to two reasons. First, this book can be only so long. Second, assessing the impact of the work of everyone who has ever lived is an impossible task, and some editorial discretion by the author is necessary.

Finally, by labeling someone a hero, this book does not commend or celebrate every aspect of that person's life. Rather, it celebrates the aspects of a person's work that helped to improve the world. With the benefit of hindsight, we can see that some of our champions held—by current standards—morally repugnant views, committed egregious acts, or were just generally unpleasant people. While their failings, moral or otherwise, should be noted and considered, they ultimately do not undermine the importance of our heroes' contributions to humanity.

What Have the Heroes of Progress Contributed?

Over the past two centuries, humanity has experienced unprecedented progress. Between 1820 and 2020, the proportion of people living in extreme poverty—defined as living on less than $1.90 per person per day—fell from 95 percent to less than 9.5 percent. This is a decline of more than 90 percentage points in just 200 years. Over the same period, average global life expectancy more than doubled from 29 years to 73 years, according to World Bank data. Also, the percentage of people above the age of 15 who are illiterate declined from 88 percent of the world's population in 1820 to just 12 percent in 2020, a decline of more than 86 percent. Similarly, as Harvard University psychologist Steven Pinker notes, "Violence has been in decline over long stretches of time, and we may be living in the most peaceful time in our species' existence."

These data will not come as a surprise to many readers interested in trends related to global well-being. For over a decade, several highly regarded authors and scholars have explored the improving state of humanity in books and on websites. (For more information, see the suggested readings at the end of this volume.)

While we as a species are becoming more prosperous, better educated, healthier, less prone to starvation, and more peaceful, it is useful to remind ourselves of the underlying cause of this progress: innovation. Human innovation—whether new ideas, inventions, or systems—is the primary way people create wealth and escape poverty. As Matt Ridley noted in his 2020 book, *How Innovation Works: And Why It Flourishes in Freedom*, innovation "is the reason most people today live lives of prosperity and wisdom compared to their ancestors, the overwhelming cause of the great enrichment of the past few centuries, the simple explanation of why the incidence of extreme poverty is in global freefall for the first time in history."

Northwestern University economic historian Joel Mokyr distinguishes between two types of economic growth. "Smithian growth," named after the founder of modern economics, Adam Smith, represents the gains caused by the expansion of trade and division of labor. "Schumpeterian growth," which is named after the Austrian economist Joseph Schumpeter, represents economic gains generated by technological progress derived from innovations. To Schumpeter, innovation drives the process he terms "creative destruction," whereby longstanding practices are replaced with more efficient techniques, products, or methods. The increased efficiencies associated with creative destruction in turn lead to long-term economic growth.

While Mokyr acknowledges the importance of Smithian growth, he believes that the immense economic progress since the industrial revolution is primarily Schumpeterian in nature. Mokyr explains, "After millennia of very slow and reversible growth, the world has taken off in the past two centuries on a path of unprecedented economic expansion, driven primarily by useful knowledge and human ingenuity." The further good news is that, as Schumpeterian growth, unlike Smithian growth, relies on new ideas leading to innovations, it can theoretically be sustained forever without diminishing returns.

However, as Ridley notes, innovation is the "most important fact about the modern world, but one of the least well understood." Considering innovation's key role in enhancing humanity's well-being, so that progress and human advancement do not cease, it is important to know where innovation comes from.

Where Does Innovation Come From?

Nations, governments, societies, businesses, committees, groups, algorithms, and computers don't have ideas. Ideas spring forth from the human brain. When people with ideas have autonomy and freedom, they turn those ideas into innovations. The human mind is unique in its ability to build on and improve past innovation, which in turn leads to greater technological progress, more economic growth, and greater human well-being.

Innovation, the engine of progress, comes from the human mind. Therefore, it is appropriate that the late University of Maryland economist Julian Simon dubbed the Homo sapiens brain "the ultimate resource." Simon is best known for arguing that by applying their intelligence and ingenuity, people can innovate their way out of shortages and make resources more abundant. More broadly, as knowledge is cumulative and therefore virtually limitless, Simon and other scholars conclude that there is no obvious reason why innovation cannot go on improving life on this planet indefinitely.

Yet despite the almost unfathomable explosion in innovation and progress of the past 200 or so years, there is nothing unique about the chemistry or makeup of our modern brain that makes us more prone to invention than our ancestors who came millennia before us. Indeed, we have always created. We gained control of fire as early as 1.7 million years ago, discovered agriculture a little over 11,000 years ago, and invented the wheel about 6,200 years ago. Historically, clusters of great transformation and relative prosperity have always occurred. Some prominent examples include the ancient Greek city-states, the Roman Nerva-Antonine Dynasty, the Abbasid Caliphate, China under the Song Dynasty, and the Dutch Golden Age. However, in nearly all previous episodes of innovation, long-term sustainable growth was not attained, progress petered out, and a "dark age" followed.

For most of human history, our relationship with innovation can best be summarized as sporadic, episodic, limited (in terms of geography, areas of study, and economic growth), and often reversible. However, in the second half of the 18th century, this relationship began to change. Starting in some western European states, humanity began a sustainable process of generating, refining, and diffusing new knowledge. Since then, abundance and prosperity have continued to reach historically unprecedented highs.

The Conditions for Innovation

What, then, differentiates our modern period of continuous innovation and prosperity from the stasis of the past? The first key factor to consider is freedom. To create, we must be free to speak, publish, associate, disagree, take risks, and disrupt the status quo. Thus, innovation is often opposed by powerful actors who benefit from keeping things as they are. A permissive social environment, in other words, is crucial if individuals are to discover, develop, and act on their ideas.

As University of Illinois at Chicago professor emerita Deirdre McCloskey notes, innovation alone "isn't enough [for technological and material progress]. It has got to be tested, whether it is profitable to work." A broad range of pro-market policies is needed to allow individuals to save, invest, trade, and ensure that the best ideas rise to the top. The policies that enable this process include low levels of taxation, secure property rights, less onerous regulations, and freedom to trade.

The price mechanism, which serves as a useful device for innovators to separate good ideas from bad, must remain undistorted for innovation to flourish. A well-functioning price mechanism generates better information, which results in a more efficient allocation of resources. Sustained innovation and progress have occurred because by increasing personal and economic freedoms, some states in western Europe—first the Netherlands and the United Kingdom—stopped disincentivizing innovation. They allowed their citizens to develop new ideas and contradict old ones without fear of ostracism, harassment, imprisonment, or death. Where western Europe led, the United States soon followed.

Although it is true that a few of our heroes hailed from or lived in tyrannical and oppressive states, that was far from the norm. In all such cases, heroes from unfree states either had to hide their work or leave their countries to avoid persecution or were awarded special privileges from the government that allowed them to continue their work. Given these conditions, the scarcity of innovation in unfree states is not surprising.

A second important ingredient in the innovation recipe is equal dignity for all people regardless of gender, race, tribe, nationality, religion, beliefs, or any other category into which people divide themselves. For maximum innovation to occur, economic and social policies should be inclusive and applied equally. If societies exclude specific groups from equal access to the market or from the education that allows people to acquire and develop new ideas, they not only are failing morally, but also are limiting the number of minds—or "ultimate resources"—that can test new ideas that could benefit everyone.

Increasing the number of human minds capable of freely dabbling in the market ties into the third key reason modernity has experienced unprecedented levels of innovation: increasing population. Since 1800, the world's population has risen from one billion people to more than eight billion. It is no coincidence that this boom in human minds has coincided with immense increases in prosperity. Having more minds means more ideas; these ideas lead to new inventions; those inventions, once tested by market forces and proven successful, turn into innovations that improve productivity, economic growth, and human well-being.

Going Forward

Ultimately, history has shown that where freedom is restricted, so too are innovation and progress. That goes a long way toward explaining the similar backgrounds of many, although certainly not all, of the people profiled in this book. Many of the world's most significant innovations described here happened between 1750 and 1970. For much of that time, social and economic freedoms and access to a good education were unfairly limited to a handful of people in just a few nations. As the world became freer, the backgrounds of the innovators became more diverse.

If, in 200 years, someone updates this book to profile the latest world-changing innovators, the nationalities, genders, religions, socioeconomic backgrounds, and creeds of those new heroes will be determined by which people had the freedom to engage in the innovative process. If every region were to become tyrannical and basic freedoms were tolerated in only a single country, most future progress would likely come from that nation. If the whole world becomes freer, then the makeup of future heroes will be more diverse.

I hope this book reminds the reader that while progress is far from guaranteed, the ability of the human mind to think its way out of problems is the most important tool we have in ensuring the prosperity of our species. For that reason and more, it is crucial that the light of freedom shine on all people and never be extinguished.

1
Johannes Gutenberg

Inventor of the printing press

Johannes Gutenberg was a 15th-century German goldsmith and inventor who created the first metal movable-type printing press. Gutenberg's inventions included a process for mass-producing movable type, the use of oil-based ink for printing books, adjustable molds, mechanical movable type, and a wooden printing press similar to the agricultural screw presses of that time. He also adapted the screw mechanism from linen-, wine-, and paper-making presses to create a machine suited to printing.

Gutenberg's ideas started a printing revolution, which greatly improved and accelerated the spread of information. The printing press helped to fuel the later part of the Renaissance, the Reformation, and the Scientific Revolution, thus setting the stage for the start of the Industrial Revolution in the second half of the 18th century.

Relatively little is known about Gutenberg's early life. It is believed he was born sometime between 1394 and 1404 in the city of Mainz, in the Holy Roman Empire (whose borders roughly corresponded to those of modern-day Germany and parts of northern Italy). We do know that Gutenberg was born into a wealthy patrician merchant family, and that he grew up learning the trade of goldsmithing.

In 1411, the Gutenbergs were exiled from Mainz following an uprising against the patrician class. We don't know much about Gutenberg's life over the following 15 years, but a letter written by him in 1434 indicates he was living in Strasbourg (which today is in France). Moreover, legal records from that same year indicate he was a goldsmith and a member of the Strasbourg militia.

While in Strasbourg, Gutenberg created metal hand mirrors that pilgrims bought and used when visiting holy sites (it was thought that mirrors could capture the holy light from religious relics). Gutenberg's metalworking skills proved useful when he developed the metal movable type used in the printing press.

In 1439, Gutenberg ran into financial problems. Unable to placate his investors, it is widely believed that he shared a secret with them. Many historians have speculated that the secret was a much-improved process of printing. A year later, Gutenberg supposedly declared that he had perfected the art of printing. That said, a workable prototype of his printing press remained a long way off.

In 1448, Gutenberg moved back to Mainz. With the help of a loan from his brother-in-law, Arnold Gelthus, he built an operating printing press in 1450. A working press enabled Gutenberg to convince Johann Fust, a wealthy financier, to lend him more capital to fund further refinement of the printing process. Peter Schöffer, Fust's son-in-law, also joined the enterprise, and it is likely that Schöffer designed some of the press's first typefaces.

It is widely accepted that Gutenberg had two presses: one for lucrative commercial texts and one reserved for the Bible. In 1455, the first 180 copies of the Gutenberg Bible were completed. However, in the same year, Fust sued Gutenberg and demanded his money back, accusing Gutenberg of misallocation of funds. The Court of the Archbishop of Mainz ruled in favor of Fust, giving him possession of the printing workshop and half of all the printed Bibles.

The court's ruling left Gutenberg effectively bankrupt. Undeterred, he managed to open a small printing shop in Bamberg, Bavaria, in 1459, where he continued to print Bibles. In 1465, the Prince-Archbishop of Mainz (who held both ecclesiastical and secular state authority) recognized Gutenberg's accomplishments by naming him a *Hofmann,* or a gentleman of the court. Until his death in 1468, Gutenberg could live comfortably on the court's large annual stipend.

Gutenberg's invention quickly spread throughout Europe and beyond. Books and pamphlets became much cheaper and more easily accessible. The deluge of printed texts helped to increase literacy rates throughout the continent. Medical, scientific, and technical knowledge proliferated, improving the lives of millions. Philosophical, religious, and political treatises abounded. In short, Gutenberg's press helped loosen the monopolistic controls that guilds and nobility had held over Europe's economic and social life for centuries.

2

Desiderius Erasmus

Champion of tolerance

Desiderius Erasmus was a 16th-century philosopher who is widely considered one of the greatest scholars of the Northern Renaissance. During the Protestant Reformation, when religious persecution was common across Europe, Erasmus advocated for religious toleration and peace. According to historian Jim Powell, senior fellow at the Cato Institute, throughout the philosopher's life, Erasmus "championed reason over superstition, tolerance over persecution and peace over war . . . [and helped to establish the] intellectual foundations for liberty in the modern world." He is often credited with being the most influential figure of the Northern Renaissance and the first modern champion of toleration and peace.

Desiderius Erasmus was born on October 28, 1466, in Rotterdam, the Netherlands. The exact year of his birth is unknown, but most scholars agree on 1466. Erasmus was the second illegitimate son of Roger Gerard, a Catholic priest, and Margaretha Rogerius. His mother was the daughter of a physician and worked as a laundress. Erasmus was educated at several monastic schools and at the age of nine was sent to one of the Netherlands' top Latin schools, in Deventer.

As a child, Erasmus witnessed a lot of religious violence. At just eight years old, he saw more than 200 prisoners of war broken on the rack following the order of a local bishop. Erasmus's early exposure to religious-based violence very likely influenced his future beliefs.

During his time in Deventer, Erasmus began to resent the harsh rules and strict methods used by his religious educators. He would later write that the severe discipline was intended to teach humility by breaking a boy's spirit. Erasmus's education in Deventer ended in 1483, when both of his parents died from the bubonic plague.

In about 1485, Erasmus and his brother were living in severe poverty. Out of desperation, they both entered monasteries. Erasmus became a canon regular of the Saint Augustine canonry in the small Dutch village of Stein. During his seven years in Stein, he spent much of his time in the library, where he studied philosophy. He was especially interested in the works of Cicero and other Roman thinkers. When his superiors became unsupportive of his classical studies, Erasmus became increasingly eager to leave the canonry.

Erasmus was ordained as a Catholic priest in April 1492. Soon after that, he left the canonry to become secretary to the Bishop of Cambrai, who had heard of Erasmus's proficiency in Latin. In 1495, Erasmus began studying theology at the University of Paris, but he grew to dislike the quasi-monastic regimen of the college.

To support his studies, Erasmus began teaching. One of his students, Sir William Blount, invited him to visit England in 1499. During Erasmus's time in England, he befriended many prominent intellectuals and began teaching at the University of Oxford. Over the next 15 years, Erasmus lived in and traveled to many places, including France, Belgium, Italy, and Switzerland. In 1509, he graduated with a doctor of divinity from Turin University, and between 1509 and 1514 he worked at the University of Cambridge. However, Erasmus was frequently in poor health and complained that Cambridge could not supply him with enough decent wine (at the time, wine was used as medicine for gallstones, from which Erasmus suffered).

In 1500, Erasmus compiled and translated 818 Latin and Greek proverbs into a volume called *Collectanea Adagiorum*. The publication of this work was the first of his many efforts to end the clergy's monopoly on learning. Throughout his life, Erasmus updated and expanded this book. By the time of his death in 1536, he had translated more than 4,151 entries. Many common phrases used today are owed to Erasmus's translation, including "one step at a time," "one to one," and "to sleep on it."

As Erasmus created more editions of *Adagia*, as the work came to be known, he added commentary advocating against violent rulers and preachers who supported wars for their own self-interest. In one passage, Erasmus questioned, "Do we not see that noble cities are erected by the people and destroyed by princes? . . . That good laws are enacted by representatives of the people and violated by kings? That the commons love peace and the monarchs foment war?"

In 1513, following the death of the "Warrior Pope" Julius II, Erasmus wrote a short satire denouncing the pope's violent ways. Erasmus implied that Julius would not get into heaven because he had spent too much time waging war rather than reading and preaching the Gospels. In 1514, Erasmus published *Familiarium colloquiorum formulae*, often referred to simply as *Colloquies.*

In *Colloquies*, Erasmus ridiculed greedy clergymen and rituals that he saw as meaningless and he declared marriage to be preferable to celibacy. Rulers in France, Spain, the Netherlands, Austria, and much of Italy decreed that anybody seen with a copy of *Colloquies* would be executed. Despite these threats, more than 24,000 copies of *Colloquies* were sold during Erasmus's lifetime. According to John Dalberg-Acton, a 19th-century historian, it was "the most popular book of its age."

In a later edition of *Adagia*, published in 1515, Erasmus also took aim at the training of preachers and argued that priests should be trained in the "philosophy of Christ" rather than in different scholastic subjects. Erasmus later argued that "if the Gospel were truly preached, the Christian people would be spared many wars."

In 1516, because of many translation errors in the official Latin Vulgate edition of the Bible, Erasmus used Greek manuscripts to produce a new, more accurate translation of the

New Testament. Erasmus's work inspired many other individuals to translate the Bible into different languages. Within 20 years, there were German, English, Hungarian, and Spanish translations. Very few people could read Latin, so these new translations helped make the Bible more accessible to people across Europe. They no longer had to rely on a preacher's interpretation of the Latin Bible and could learn about religion on their own.

In the following years, Erasmus released a series of works making the case for peace and proclaiming that salvation was achieved not by performing religious rituals, but by cultivating faith and goodness. Contrary to many thinkers of his day, Erasmus also argued against colonialism and for a "limited monarchy, checked and decreased by an aristocracy and by democracy."

Erasmus considered himself an impartial party during the Protestant Reformation, criticizing both the Catholic church hierarchy and the Protestant reformers. He remained a Catholic, committed to reforming the church from within. Throughout his life, Erasmus had many arguments with Martin Luther, a leading Protestant figure who helped launch the Reformation.

Erasmus claimed that Luther was an enemy of liberty and that he was on the side of tyrannical rulers rather than the people. Toward the end of his life, Erasmus warned that religious wars could soon break out in Europe. In one of his last works, *On the Sweet Concord of the Church,* Erasmus made one final plea for Catholics and Protestants to "tolerate one another." Erasmus died a painful death on July 11, 1536, in Basel, Switzerland, following a three-week struggle with dysentery. He was buried in Basel Cathedral.

Unfortunately for the people of Europe, Erasmus's predictions of religious wars came true. In the decades after his death, Europe was embroiled in religious violence that saw hundreds of thousands of Catholics and Protestants slaughter one another. However, a couple of hundred years after his death, a newfound appreciation for Erasmus's work emerged among Enlightenment scholars. French dramatist and encyclopedist Denis Diderot, for example, noted, "We are indebted to him, principally for the rebirth of the sciences, criticism, and the taste for antiquity." Today, Erasmus has been criticized for holding anti-Semitic views (which were common in his day), due mostly to his abhorrent characterization of Jews as "the most noxious pests."

Throughout his life, Erasmus was offered reputable positions in academic institutions across Europe, but he declined them all, preferring a life as an independent scholar dedicated to writing works that would help societal progress. Although Erasmus's works failed to prevent Europe's religious wars, his writings went on to have enormous influence on the Enlightenment's ideas of rationality, peace, and liberty.

3
Sir John Harington

Inventor of the flushable toilet

Sir John Harington was a 16th-century English courtier, author, and inventor of the modern flush toilet. Harington's toilet allowed waste to be flushed from people's houses to underground cesspools without direct human contact. The flushing toilet has had immeasurable sanitary benefits for the modern world—the World Economic Forum has estimated that its invention has helped save more than one billion lives.

John Harington was born into a wealthy noble family on August 4, 1560, in Kelston, a town in southwest England. Upon his christening in London a few months later, he became one of Queen Elizabeth I's 102 godchildren. Harington's father, also called John, was a poet at the court of Henry VII, and his mother, Isabella Markham, was a gentlewoman in Queen Elizabeth I's privy chamber. Harington was educated at Eton College, an elite all-boys boarding school, before studying law at King's College, Cambridge.

While his father wanted him to become a lawyer, Harington became enamored with life at the royal court. His propensity to speak freely quickly gained him notoriety among the nobles. Queen Elizabeth was fond of Harington and often encouraged him to write poetry. However, Elizabeth would come to regret that impulse, as Harington became known for writing risqué poems and epigrams that often overstepped what was deemed morally permissible at court.

Harington's first banishment from court resulted from an escapade in 1584, when he translated chapter 28 of Ludovico Ariosto's epic poem *Orlando Furioso*. He circulated the manuscript among the court's maids of honor. Angered by the racy translation, Elizabeth exiled Harington and told him that he would not be allowed to return to court until he had translated all 40 chapters of *Orlando Furioso*—a task so arduous that many assumed Harington would fail.

However, Harington completed the full translation of the poem eight years after his banishment in 1592 and presented Elizabeth with a bound copy of the work when she visited Kelston that year. Harington's translation was widely praised and is still read today. It was during his time in exile from court that Harington designed and installed the first flushing lavatory, which he dubbed Ajax ("jakes" is an old slang word for toilet) at his Kelston manor.

Harington's device had a seat and a pan, with an opening at the bottom that was sealed with a leather-capped valve. Levers and weights poured water from a cistern above into the toilet. When the handle of the seat was turned, a valve at the bottom of the pan opened and water swept the pan's contents into a cesspool below. Harington first described his invention in his 1596 book *A New Discourse upon a Stale Subject: The Metamorphosis of Ajax*, which he published under the pseudonym Misacmos, meaning "hater of filth." In his book, Harington declared that his Ajax toilet could ensure that "unsavoury places may be made sweet, noisome places made wholesome, [and] filthy places made cleanly."

In his book, Harington, never one to miss out on an opportunity to make a political statement, made numerous digressions, often aimed at well-known men at court. The book was in large part an attack on the supposed "excrement" that was poisoning society and contained many allusions to Queen Elizabeth's favorite, the Earl of Leicester. Although his book enjoyed considerable popularity, Harington was threatened with a hearing in front of the Star Chamber, an English court in the Palace of Westminster. Although Elizabeth's fondness for Harington protected the inventor from more severe punishment, he was once again banished from the royal court.

In 1598, Elizabeth asked Harington to install a toilet at Richmond Palace, a royal residence on the river Thames. The toilet became popular among members of the nobility, but much of the public remained faithful to their chamber pots. It wasn't until almost 200 years later that Scottish inventor Alexander Cumming patented the flushing water closet inspired by Harington's Ajax. Cumming's 1775 design improved on Harington's device by adding the S-trap in the piping below the toilet, which allowed water to be permanently retained in the pipe, thus preventing sewer gases from entering the buildings above.

In 1848, a Public Health Act in the United Kingdom ruled that every new house required a "w.c., privy, or ashpit." It took over 250 years for Harington's flushing toilet to catch on among the general public. Today, more than two-thirds of the world's population has access to a flushing toilet, and that figure continues to rise by tens of millions every year.

In 1599, Harington joined an English military campaign in Ireland to subdue a rebellion by Gaelic chieftains. He was knighted for his service. After Harington's time in Ireland, he tutored James I's son Henry, Prince of Wales. Harington died on November 20, 1612, at his home in Kelston. He was 52 years old.

Toilets fundamentally changed the world in which we live. The sanitary benefit of not having to be in direct contact with human waste prevents millions of cases of cholera, diarrhea, dysentery, hepatitis A, typhoid fever, and polio every year.

4
Mary Astell and Mary Wollstonecraft

Advocates of women's equality

Mary Astell and Mary Wollstonecraft were 17th- and 18th-century English thinkers who are widely considered to have been the pioneers of feminist philosophy. Their works gained popularity in the 19th century and helped to provide the philosophical foundation for suffragette and women's rights movements all over the world.

In the 17th and 18th centuries, western European women were often poorly educated and had very little protection under the law. In a series of prominent works, Astell argued that women should have the same educational opportunities as men. Astell was also the first Englishwoman to base her justifications for gender equality in philosophy rather than on historical evidence.

Wollstonecraft took Astell's call for equal education for the sexes a step further. She contended that because both men and women were endowed with the inalienable rights to life, liberty, and the pursuit of happiness, women should also be granted the right to vote and be allowed to pursue whatever career they wished.

Mary Astell was born into an upper-middle-class family in Newcastle, England, on November 12, 1666. Her father managed a local coal company, but despite her family's wealth, Astell did not receive any formal education. Instead, she was taught at home by her clergyman uncle, Ralph Astell. Ralph was heavily involved in the philosophical school called Neoplatonism, which taught rationalist beliefs based on the works of the Greek philosophers, such as Plato, Aristotle, and Pythagoras.

At the age of 12, Astell's father died and left her without a dowry, which meant that her prospects for marrying someone of a similar social class became slim. A year later, her uncle Ralph died, leaving her without a teacher. Yet throughout her teenage years, Astell continued to teach herself many subjects and discovered that she had a particular aptitude for political philosophy.

In 1684, Astell's mother died. That spurred Astell to move to Chelsea, a suburb of London, where she quickly became acquainted with a circle of literary and influential women. Astell's new friends, along with William Sancroft, former Archbishop of Canterbury, who provided her with financial support, helped Astell develop and publish her writings.

In 1694, Astell published her first book, *A Serious Proposal to the Ladies, for the Advancement of Their True and Greatest Interest.* Six years later, her second book, *Some Reflections upon Marriage*, was published. In both works, which were published anonymously, Astell argued that women should receive an education equal to that of men. She averred that the existing intellectual disparity between men and women was due not to a natural inferiority, but to the latter's lack of educational opportunity. Astell also declared that women should be able to choose whom to marry or to refrain from marriage if they so desired.

Through these works, Astell became one of the first writers to advocate for the idea that women were just as rational as men. By using French philosopher René Descartes's theory of dualism—the idea that the mind and body are distinct and separable—Astell argued that the genders had equal abilities to reason, irrespective of their physical differences. Therefore, women should be treated as equals. Astell famously wrote, "If all men are born free, why are all women born slaves?"

Later in life, Astell left the public eye. In 1709, she became head of a charity school for girls. She designed the curriculum, and it may have been the first school in England to have an all-women board of governors. After a mastectomy to remove a cancerous breast, Astell died at her Chelsea home on May 11, 1731.

Throughout her life, Astell encouraged both women and men to fight for women's rights. After her death, Mary Wollstonecraft continued the advocacy for educational reform for women that Astell had begun.

Mary Wollstonecraft was born on April 27, 1759, in London, England. Like Astell, she was born into an upper-middle-class family that became significantly poorer over time. Wollstonecraft's father, Edward John Wollstonecraft, was a violent man who frequently beat his wife in drunken rages. As a child, Mary would often intervene and try to prevent her father's abuse. Edward Wollstonecraft gradually squandered the family's money, causing the family to move several times during her childhood.

Early in Wollstonecraft's life, she befriended Jane Gardiner, née Arden. The pair would often read then-new Enlightenment-era books together. They also attended lectures by Gardiner's father, John Arden, a scholar of natural philosophy who was one of Wollstonecraft's early teachers.

Unhappy with her home life, Wollstonecraft decided to move. Throughout the late 1770s and early 1780s, she worked at several different jobs across England and Ireland, including as a governess, needleworker, and teacher.

Wollstonecraft became frustrated with the limited career options open to women. In the late 1780s, she embarked on a career as an author, which was seen as a radical choice for a woman at the time. In 1787, Wollstonecraft wrote her first book, *Thoughts on the Education of Daughters*. The work resembles an early self-help book and offers advice on women's education. It also includes segments on morality, etiquette, and basic child rearing.

In 1788, Wollstonecraft was employed as a French and German translator for London publisher Joseph Johnson, who published several of her early works. Wollstonecraft had a great interest in the French Revolution. After the Anglo-Irish philosopher and member of Parliament Edmund Burke published the book *Reflections on the Revolution in France*, which challenged the principles of the French Revolution, Wollstonecraft decided to respond.

In 1790, Wollstonecraft published *A Vindication of the Rights of Men*, which criticized the arbitrary nature of government power and the despotism of France's *Ancien Régime*, welcomed revolutionary reform, and argued that humanity's natural rights must be protected by governments.

In 1792, Wollstonecraft published her best-known work, *A Vindication of the Rights of Woman.* In the book, Wollstonecraft expanded on Astell's work and argued that the education system trained women to be frivolous and incapable. Wollstonecraft noted that there are no mental differences between men and women. If women were given the same educational opportunities as men, she contended, they would be capable of doing many professions and elevating themselves in society.

Unlike Astell, Wollstonecraft believed that the betterment of women should be achieved through radical political change, with reforms required in both the educational and the voting systems. Wollstonecraft stated that, as men and women are intellectually similar, women should also be granted the right to vote. She wrote, "Women ought to have representatives, instead of being arbitrarily governed without having any direct share allowed them in the deliberations of government."

Wollstonecraft also declared, "Liberty is the mother of virtue." If women were kept "by their very constitution, slaves, and not allowed to breathe the sharp invigorating air of freedom, they must ever languish like exotics, and be reckoned beautiful flaws of nature."

A Vindication of the Rights of Woman was hugely successful and helped bolster Wollstonecraft's reputation as a writer. Later in 1792, Wollstonecraft went to Paris to observe the French Revolution. She arrived just a month before King Louis XVI was guillotined. Wollstonecraft stayed in France until 1795.

After the breakdown of a romantic relationship, Wollstonecraft was left heartbroken and twice attempted suicide. When she returned to England, she became involved in a close-knit group of radical intellectuals, which included William Godwin, Thomas Paine, William Blake, and William Wordsworth.

In 1797, Wollstonecraft married William Godwin and gave birth to Mary Wollstonecraft Shelley, who would go on to write *Frankenstein.* On September 10, 1797, just 11 days after she gave birth, Wollstonecraft died of septicemia.

The writings of Astell and Wollstonecraft were unsuccessful in bringing about immediate reforms when they were published. The works, however, provided the intellectual foundation for the suffragette and feminist movements, which began in the late 19th century and continue around the world. Although these two women were considered radicals during their time, without their ideas, it is unlikely that women's rights would be as extensive as they are today.

5

James Watt

Father of the Industrial Revolution

James Watt was an 18th-century Scottish engineer and inventor who enhanced the design of the steam engine. Watt's steam engine made energy supply more efficient and reliable than ever before and it was fundamental to kick-starting the Industrial Revolution.

James Watt was born on January 19, 1736, in Renfrewshire, Scotland. His father was a successful shipbuilder, and Watt later reminisced that growing up around his father's workshop had a profound influence on his educational goals and career trajectory. Because of bouts of illness as a child, Watt was mostly homeschooled.

In 1755, when he was 18, Watt's mother passed away, and the future inventor traveled to London to study mathematical instrument making, which involved learning to build and repair devices such as quadrants, compasses, and scales. After a year in London, he returned to Scotland, where he produced and repaired mathematical instruments. Watt eventually opened a mathematical instrument shop in 1757 at the University of Glasgow.

In 1764, the university gave Watt a Newcomen steam engine to repair at his workshop. This was an older engine that had been invented in 1712. The Newcomen engine operated by condensing steam in a cylinder, which creates enough pressure to power a piston. While fixing the engine, he observed that a lot of the steam was wasted because of the machine's single-cylinder design. He found that having to heat and cool the same cylinder repeatedly wasted more than three-quarters of the thermal energy built up by cooling the steam.

To remedy this inefficiency, in 1765 Watt created a design that used steam condensed in a chamber that was separated from the cylinder. This design was revolutionary. Unlike the Newcomen engine, which squandered energy by heating and cooling the same cylinder over and over, Watt's engine kept the cylinder at a stable temperature as the steam condensed in a separate chamber.

However, due to a lack of capital, Watt faced difficulties in constructing a full-scale engine. With an investment from Joseph Black, a University of Glasgow physician, Watt built a small test engine in 1766. A year later, Watt entered a business partnership with John Roebuck. In 1769, Watt and Roebuck took out their famous patent for "A New Invented Method of Lessening the Consumption of Steam and Fuel in Fire Engines."

Unfortunately, acquiring the patent bled Watt's funds dry, and he was forced to take on alternate employment, first as a surveyor and then as a civil engineer.

Seven years later, Watt's old business partner went bankrupt, and an English manufacturer named Matthew Boulton acquired Roebuck's patent rights. Thanks to Boulton, Watt returned to working full-time on his engine.

Together, the men founded the Boulton and Watt manufacturing firm, in which Watt spent the next several years improving the efficiency and cost of his engine. Watt's first profitable dual-cylinder steam engine came on the market on March 8, 1776, a day before Adam Smith's *Wealth of Nations* was published. Little did the two Scotsmen know that they were about to change the world forever.

As demand for Watt's engine grew, it was rapidly adopted across multiple industries, including in rotary machines that were used in cotton mills, which enabled cheap clothing to be made for the masses for the first time. Ultimately, Watt's design turned the steam engine from a machine that yielded marginal efficiency into a tool that powered much of the Industrial Revolution and therefore helped create the modern world we know.

In 1800, when the patent on the steam engine expired, Watt retired. On August 15, 1819, Watt died at the age of 83 in Birmingham, England. He was honored with numerous awards during his lifetime, including fellowships of the Royal Societies of both London and Edinburgh. In 1960, the watt unit of power was named after him. In 2009, the Bank of England put Watt's face on the new £50 note.

Industrialization has enabled hundreds of millions of people to rise out of poverty. Today, all developed countries have gone through the process of industrialization, a phenomenon that could not have occurred without the Watt engine.

6

Cesare Beccaria

Father of criminal justice

Cesare Beccaria, an 18th-century Italian criminologist, was the first modern writer to advocate for the abolition of capital punishment and the end of cruel, torturous punishments. Many consider Beccaria to be the father of criminal justice. He believed that penalties for crimes should be proportional to the severity of the offense and that criminals should not be punished until proven guilty in a court of law. Thanks to his work, many nations were inspired to enact extensive legislative reforms to ensure due process and the end of torture and the death penalty.

Cesare Beccaria was born March 15, 1738, in Milan, Italy. His father was an aristocrat on a moderate income. At the age of eight, Beccaria was sent to a Jesuit boarding school in Parma.

He excelled in mathematics, although his early student days gave little indication of his intellectual brilliance. As a child, Beccaria was prone to angry outbursts, which caused him periods of immense enthusiasm, followed by periods of depression and inactivity—a pattern that would affect him for the rest of his life.

In 1754, Beccaria enrolled at the University of Pavia, in the Italian region of Lombardy, and received his law degree in 1758. In his mid-twenties, Beccaria became friends with Pietro and Alessandro Verri, brothers and writers from the Milanese aristocracy. Together, the young men formed a literary society they called the Academy of Fists. The playfully named group dedicated itself to the promotion of economic, political, and administrative reforms. The society read many of the French and British Enlightenment thinkers and published a magazine named *Il Caffè*. The magazine, modeled on the English *Spectator*, sought to introduce Italians to Enlightenment ideas.

In 1763, inspired by his involvement in the Academy of Fists, Beccaria turned his attention to the study of criminal law. Although he had no prior experience working on criminal justice, in 1764 he published his most influential essay, "On Crimes and Punishments."

The short essay heavily critiqued the use of torture, the arbitrary discretionary power of judges, the lack of consistency and equality of sentencing, and capital punishment. Beccaria argued that sentences should be scaled to the severity of an offense and should be only harsh enough to ensure security and order. He wrote that anything beyond that would be tyranny. The goal of Beccaria's essay was to critique the existing legal system, which he felt was unclear, imprecise, and based largely on a mixture of Roman law and local customs rather than on rationality. Beccaria believed that the government deliberately exploited the opacity of the laws to control the populace.

Beccaria's essay argued that the effectiveness of criminal justice depends mostly on the certainty of punishment rather than its severity. Unlike many works before it, Beccaria's publication also advocated that no one should be sentenced until proven guilty in a court of law.

Beccaria's essay became a success and was quickly translated into French, English, Dutch, German, and Spanish. Initially, Beccaria published the essay anonymously, fearing government reprisals. However, after its rapid success, he soon republished it and credited himself as the author. Soon after its publication, Empress Catherine the Great of Russia publicly endorsed Beccaria's ideas, and Thomas Jefferson and John Adams later noted Beccaria's influence on the American Constitution and the Bill of Rights. Legislative reforms in Sweden, Russia, and the Habsburg Empire were heavily influenced by Beccaria's treatise, and the essay exerted enormous influence on criminal law reforms across other parts of the European continent.

In the late 1760s, Beccaria turned his attention to the study of economics, although none of his later works achieved the same success as "On Crimes and Punishments." In 1768, he accepted the chair in public economy and commerce at the Palatine School in Milan. Two years later, Beccaria was appointed to the Supreme Economic Council of Milan. While in office, he focused mainly on the issues of public education and labor policy. One of his later reports also played an important role in influencing France's subsequent adoption of the metric system.

Beccaria's later life was marred by family difficulties and health problems. Property disputes between his siblings resulted in litigation, which distracted him for many years. He cut short his visit to Paris in 1766 because of homesickness and never ventured abroad again. Over many years, a myth grew that Beccaria's literary silence was due to his expulsion from the Milanese government. In reality, his silence was caused by periodic bouts of depression. Although Beccaria was initially enthusiastic about the French Revolution, he spent his last few months saddened by the violence of the Reign of Terror. He died on November 28, 1794, in Milan.

Beccaria's work fundamentally changed the criminal justice systems in many countries for the better. As the first modern writer to champion the abolition of the death penalty, he can be regarded as the founder of the anti-capital punishment movement that still exists in many countries. Thanks to Beccaria, cruel and unusual punishments are no longer the norm in much of the world. As a result of his advocacy, an immense amount of human suffering and injustice has been avoided.

7

Alessandro Volta

Inventor of the electric battery

Alessandro Volta was an 18th-century Italian physicist who invented the electric battery. His "voltaic pile" provided the first source of continuous electric current the world had ever seen. Through his discovery, Volta debunked the theory (prevalent at the time) that electricity was generated solely by living beings. Volta's invention laid the groundwork for modern batteries. His work also helped to create the fields of electrochemistry and electromagnetism.

Alessandro Giuseppe Antonio Anastasio Volta was born on February 18, 1745, in Como, a town in present-day northern Italy. Volta's family was noble and wealthy. As a child, he attended a Jesuit boarding school, where his teachers tried to persuade him to enter the priesthood. Volta knew that his passion lay in physics, and at the age of 16, he dropped out of school. Despite not receiving any further formal schooling, Volta began to exchange letters with leading physicists of the day by the time he was 18. Two years later, he was already conducting experiments in a physics lab built by his wealthy friend Giulio Cesare.

By 1774, Volta was teaching experimental physics in Como's public grammar school. At this point, his work focused primarily on the chemistry of gases. In 1778, after reading a paper by Benjamin Franklin on the topic of "flammable air," Volta became the first person to isolate the gas methane. He found that a methane-air mixture could explode with an electric spark when in a closed container. This type of electrically induced chemical reaction would later become the basis of the internal combustion engine.

In 1779, Volta was appointed professor of experimental physics at the University of Pavia, a position he would maintain for almost 40 years. Volta spent his first years in Pavia studying what we now call "electrical capacitance." He found that electrical potential in a capacitor—a component with the capacity to store energy in the form of an electrical charge—is directly proportional to its electric charge. Today, this phenomenon is known as Volta's Law of Capacitance.

In 1791, Volta's friend and fellow physicist Luigi Galvani found that he could get a frog's leg that was mounted on iron or brass hooks to twitch when it was touched with a probe made from another metal. Galvani interpreted his discovery as a new form of electricity that can be found in living tissue and named it "animal electricity." Volta disagreed with Galvani's findings.

He hypothesized that the frog merely conducted the electrical current, which flowed between the iron or brass hook and the other metal that was being used as a probe. Volta called this type of electricity "metallic electricity."

Volta then began experimenting to see if he could produce an electrical current with metals alone. As instruments at the time were unable to detect weak electrical currents, Volta tested the flow of electricity between different metals by placing them on his tongue. Sure enough, Volta found that the saliva in his mouth, like the frog's tissue in Galvani's experiments, conducted electricity, causing a bitter sensation.

To show conclusively that an electric current did not require the presence of animal tissue, Volta created a stack of alternating zinc and silver discs, which were separated by brine-soaked cloth. He found that when a wire was connected to both ends of the pile, a steady current flowed between the layers. This invention, which came to be known as the voltaic pile, was an early form of the electric battery. After numerous experiments, Volta also found that the amount of current produced could be increased or decreased by using different metals or adding and taking away disks from the pile.

Volta first reported his electric pile experiment in a letter dated March 20, 1800. It was addressed to Joseph Banks, the president of the Royal Society of London. Soon after, Volta traveled to Paris to demonstrate his invention, which he initially called an "artificial electric organ."

Volta's battery was a huge success. Not only did it undermine the scientific consensus around animal electricity, but the artificial electric organ was also quickly recognized as an extremely useful device. Within six weeks of Volta's announcement, English scientists William Nicholson and Anthony Carlisle used their own voltaic pile to decompose water into hydrogen and oxygen. This led to the discovery of electrolysis, a method of passing an electric current through a substance to drive a chemical reaction. With that discovery, Nicholson and Carlisle helped lay the foundations for the field of electrochemistry, the study of chemical processes that cause electrons to move. In the 1830s, another English scientist, Michael Faraday, used the voltaic pile in his groundbreaking studies of electromagnetism.

Napoleon Bonaparte was so impressed with Volta's work that in 1801, he made Volta a count and senator of the kingdom of Lombardy. In 1809, Volta also became an associated member of the Royal Institute of the Netherlands.

Volta retired in 1819 at the age of 74. He moved to his estate in Camnago, which was later renamed Camnago Volta in his honor. On March 5, 1827, Volta died at the age of 82. Since his death, Volta has appeared on Italian postage stamps and on the nation's former currency, the lira (before the euro was adopted in 1999). His name was immortalized when the measure of electric potential, or volt, was named after him in 1881.

Volta's invention of the battery helped lay the groundwork for the creation of several scientific fields. In addition, the battery has become a staple of the modern world. Without Volta's work, many of our modern technologies would not exist.

8
Jeremy Bentham

Utilitarian philosopher and reformer

Jeremy Bentham was an 18th-century English philosopher, Enlightenment thinker, and social reformer. He is regarded as the founder of utilitarianism, a philosophy that holds that the most ethical choice in any situation is the one that will produce the greatest good for the largest number of people. Throughout his life, Bentham advocated for many things, including the separation of church and state, individual and economic freedoms, women's suffrage, the right to divorce, decriminalization of homosexuality, and freedom of expression. He is also widely regarded as one of the earliest proponents of animal rights.

Although the impact of Bentham's work during his lifetime was relatively small, he inspired countless other thinkers and politicians who were successful in implementing an array of important social reforms.

Jeremy Bentham was born into a wealthy family on February 15, 1748, in London, England. Reportedly, he was a child prodigy. At the age of three, he began to study Latin. By the age of seven, he was a gifted violin player and would regularly perform Handel sonatas at dinner parties. As a child, Bentham attended the elite Westminster School. At the age of 12, he went to study law at the University of Oxford.

Bentham completed his bachelor's degree in 1763 at the age of 15. He completed his master's degree three years later. Although he was called to the bar in 1769, Bentham never practiced law. During his time at Oxford, he found he had little regard for the complexities of English law. Instead, he decided to spend his time trying to reform it.

Thanks to his family's wealth, Bentham could devote himself to intellectual pursuits full-time. Bentham's first book, titled *Fragment on Government,* was published in 1776. The book concentrated largely on rejecting Sir William Blackstone's work *Commentaries on the Laws of England,* which aimed to set out the legal basis of the British constitution. Bentham rejected the famed jurist's treatises, averring that England should "break loose from the trammels of authority and ancestor-wisdom on the field of law." Instead, Bentham argued, the law should be based on the principle of utility.

Bentham once stated that the inequality of men and women under the law made him choose a career as a reformer at an early age. His works often advocated for complete equality of the sexes. In 1785, Bentham argued for abolishing laws that prohibited homosexuality. Although the essay remained unpublished during Bentham's lifetime (likely out of fear of offending public morality and provoking government reprisals), it remains one of the first arguments in favor of legalization of private same-sex relations.

On a visit to Russia in 1785, Bentham wrote his first essay on economics, titled "Defence of Usury." The essay shows that Bentham held beliefs similar to those of Scottish economist Adam Smith, who is commonly known as the father of classical economics. In his treatise, Bentham argued against a planned economy and declared that each individual was the best judge of their own advantage. However, he differed from Smith in his belief that interest rates should be allowed to float freely rather than be set by a governing body. Bentham's later works on political philosophy largely followed a laissez faire approach.

In 1789, Bentham published one of his most successful books, *An Introduction to the Principles of Morals and Legislation.* The publication is often considered his most important theoretical work. In that work, Bentham developed his theory of utilitarianism. He suggested that humanity was governed by two primary motivations: pain and pleasure. The objective of all legislation, therefore, should be to ensure the "greatest happiness of the greatest number." Having been inspired by Cesare Beccaria (see chapter 6), Bentham thought punishment is justifiable only "so far as it promises to exclude some greater evil."

Bentham was raised in a Tory household, but his views on representative government departed from traditional conservatism. His 1809 paper, "A Catechism of Parliamentary Reform," advocated for annual elections, the secret ballot, wider suffrage that included women, and the freedom for women to participate in government.

In 1823, Bentham cofounded *The Westminster Review* with his student James Mill, father of the important classical liberal philosopher John Stuart Mill (see chapter 14). The publication was described as a journal for "philosophical radicals." This group of Bentham's disciples went on to have a significant influence in British public life and, by extension, on the legal foundations of most common-law nations today, including Australia and Canada.

In one of his later works, which was published as a letter to the editor of the *Morning Chronicle* in 1825, Bentham embraced the cause of animal rights. Bentham professed that, regarding mindless torture or inflicting of pain, the "insuperable line" should not be the ability to reason, but the ability to suffer. He argued that if the ability to reason was the sole criterion by which rights were attributed, then human infants or people with certain disabilities might fall short. However, Bentham did make it clear that animals could be killed for food or in defense of human life, provided that the animal was not made to suffer unnecessarily.

On June 6, 1832, Bentham passed away at his residence in Westminster, London. He was 84. Bentham continued writing up to a month before his death. He left detailed instructions for his body to be dissected and then carefully preserved and displayed. Bentham's wishes were fulfilled; to this day, his mummified body remains on display in the Student Centre at University College London.

Jeremy Bentham was one of the most important figures of the British Enlightenment. Although his works did not lead to legislative reforms during his lifetime, many of Bentham's ideas had an enormous impact on shaping the development of rational, utility-based laws in many countries. Bentham's work has helped to shape a more humane world.

9

Edward Jenner

Inventor of vaccination

Edward Jenner was an 18th-century English physician and surgeon who pioneered the smallpox vaccination—the world's first vaccine. For his work, Jenner is often dubbed the father of immunology.

Before it was eradicated in 1979, smallpox was one of humanity's oldest and most devastating scourges. The virus, which can be traced back to Pharaonic Egypt, is thought to have killed between 300 and 500 million people in the 20th century alone.

Smallpox, or the "speckled monster," as it was known in 18th-century England, was highly contagious and left the victim's body covered with abscesses that caused serious scarring.

If the viral infection was strong enough, the patient's immune system collapsed, and the person died. The mortality rate for smallpox was between 20 and 60 percent, and of those lucky enough to survive, a third were left blind.

Edward Jenner was born on May 17, 1749, in Gloucestershire, England. He was the eighth of nine children, and his father, Reverend Stephen Jenner, was a local vicar. Following his father's death when he was five, Edward was raised primarily by an older brother who was also in the clergy.

Jenner was inoculated against smallpox by variolation when he was eight. Variolation was an earlier way to immunize people. Most commonly, it involved inserting or rubbing powdered smallpox scabs from an infected person into a scratch on a healthy person. Although variolation was often successful, it was much more dangerous than later vaccines as a way to immunize people against smallpox. It often caused health problems and occasionally failed to protect individuals against smallpox later in life. Although Jenner's variolation did protect him from smallpox later in life, the procedure had a lifelong effect on his health.

As a child, Jenner attended a local grammar school, and at the age of 13 or 14, he became an apprentice to a nearby surgeon. In 1770, at 21, Jenner enrolled at St. George's Hospital in London to complete his medical training. At St. George's, he became a pupil of John Hunter, one of the most prominent surgeons in England at the time, and the pair became lifelong friends.

At the hospital, Jenner had various interests. He studied geology and clinical surgery, conducted experiments on human blood, built and twice launched his own hydrogen balloons, and conducted a lengthy study on the cuckoo. During his studies,

Jenner was offered the post of naturalist on board Captain James Cook's second expedition to the South Seas. Jenner refused the offer, and by 1773, he had returned to Gloucestershire to work as a local practitioner and surgeon.

In 1796, Jenner turned his attention to smallpox. For many years he had heard stories that dairymaids (women who work milking cows and making cheese and butter) were immune to smallpox because they had already contracted cowpox, a mild disease from cows that resembles smallpox, when they were children. If this rumor were true, Jenner wondered if cowpox could be deliberately transmitted to people to protect them against smallpox.

Jenner found a young dairymaid named Sarah Nelmes, who had recently been infected with cowpox from Blossom, a cow whose hide still hangs on the wall of St. George's Hospital. On May 14, 1796, he extracted pus from one of Nelmes's pustules and—quite unethically by today's standards—inserted it in an eight-year-old boy named James Phipps, the son of Jenner's gardener.

Over the next two weeks, Phipps developed a mild fever but no visible infection. Almost two months later, on July 1, 1796, Jenner inoculated the boy with matter from a fresh smallpox lesion, and miraculously, no disease developed. Jenner concluded that the experiment had been a success, and he named the new procedure "vaccination" from *vacca*, the Latin word for cow.

American physician Donald Hopkins has noted, "Jenner's unique contribution was not that he inoculated a few persons with cowpox, but that he then proved that they were immune to smallpox."

At first, the medical profession was hostile to Jenner's method. After vaccinating a few more patients, in 1798 Jenner published his findings in a book titled *An Inquiry into the Causes and Effects of the Variolae Vaccinae.* Over time, he successfully convinced some doctors to test the procedure. Once the merits of Jenner's vaccination became evident, the procedure quickly spread around Europe.

Jenner soon received international acclaim. Empress Catherine II of Russia gifted him a ring, Napoleon Bonaparte minted a unique medal for him in 1804, and the chiefs of the Five Nations of the Iroquois (Mohawk, Onondaga, Seneca, Oneida, and Cayuga) sent him a string and belt of wampum beads to express their gratitude. Napoleon, who was at war with Britain at the time, had all his troops vaccinated and even released two English prisoners at Jenner's request. Napoleon is cited as saying he could not "refuse anything to one of the greatest benefactors of mankind."

Jenner made no attempt to enrich himself through his discovery and even built a small one-room hut in his garden, where he would vaccinate the poor free of charge—he called it the "Temple of Vaccinia." In fact, he devoted so much of his time to providing free vaccinations that his finances suffered. In 1802 and 1806, Parliament voted to give him £10,000 and £20,000 (worth about $1 million and $2 million, respectively, in 2023 U.S. dollars).

Later in life, he was appointed Physician Extraordinary to King George IV and was made mayor of Berkeley, Gloucestershire. He continued to study natural history, and in 1823, he presented *Some Observations on the Migration of Birds* to the Royal Society. A few weeks later, on January 26, 1823, he died of an apparent stroke.

In 1980, a few months after a group of independent scientists certified its eradication, the 155 member states of the World Health Organization officially declared smallpox an eradicated disease. The smallpox vaccine laid the foundation for other discoveries in immunology and the amelioration of diseases such as rubeola (measles), influenza (the flu), tuberculosis, diphtheria, tetanus (lockjaw), pertussis (whooping cough), hepatitis A and B, polio, yellow fever, rotavirus, and COVID-19.

Jenner's work has saved untold hundreds of millions of lives not only from smallpox, but also from dozens of other diseases that had plagued humanity for millennia.

10
James Madison

Father of the U.S. Constitution

James Madison was a Founding Father of the United States and the country's fourth president. He composed the first drafts and the basic frameworks for the U.S. Constitution and the Bill of Rights. Madison, often called the Father of the Constitution, spent much of his life ensuring that the U.S. Constitution was ratified and that the freedoms of religion, speech, and the press were protected under the law.

James Madison was born on March 16, 1751, in Port Conway, Virginia, and was raised on his family's plantation. His father was one of the largest landowners in the Piedmont area. Although Madison was the oldest of 12 children, only six of his siblings would live to adulthood (a common occurrence at that time, even among the wealthy). In the early 1760s, the Madison family moved to the Montpelier estate in Virginia.

As a teenager, Madison studied under several well-known tutors. Unlike most wealthy Virginians of his day, he did not attend the College of William and Mary. Instead, in 1769 Madison enrolled at the College of New Jersey, now Princeton University, which he chose primarily for its tolerance of anti-clerical views. Despite being an Anglican, Madison was opposed to an American episcopate. He saw it as a way of strengthening the power of the British monarchy and as a threat to the colonists' civil and religious freedoms.

At the College of New Jersey, Madison completed his four-year course in just two years. After graduating in 1771, Madison remained in New Jersey to study Hebrew and political philosophy under the college's president, and future Founding Father, John Witherspoon. Madison's thoughts on philosophy and morality were strongly influenced by Witherspoon. Terence Ball, a biographer of Madison, noted that in New Jersey, Madison "was immersed in the liberalism of the Enlightenment and converted to eighteenth-century political radicalism."

In 1773, Madison returned to Montpelier. Without a career, he began to study law books and took an interest in the relationship between the American colonies and Great Britain. In 1775, when Virginia began preparing for the Revolutionary War, Madison was appointed as a colonel in the Orange County militia. However, he was frequently in poor health. Madison never saw battle and soon gave up on a military career. Instead, he entered politics. In 1776, Madison represented Orange County at the Virginia Constitutional Convention, where he helped design a new state government, independent from British rule.

During his time at the Virginia Constitutional Convention, Madison fought for religious freedom and was successful in convincing delegates to amend the Virginia Declaration of Rights to provide for "equal entitlement" rather than just "tolerance" in the exercise of religion. While at the convention, he met Thomas Jefferson, a fellow Founding Father, who became the third president of the United States and his lifelong friend.

Following the enactment of the Virginia Constitution in 1776, Madison was elected to the Virginia House of Delegates and soon after to the Council of State for Virginia's governor, who was then Jefferson. In 1780, Madison traveled to Philadelphia as a Virginia delegate to the Continental Congress, a body of delegates from the 13 American colonies that would create the United States of America.

The Articles of Confederation were ratified by the Constitutional Congress in 1781 and served as the first constitution for the 13 former colonies. The Articles gave great powers to the states, which acted more like individual countries than a union. Madison felt that this structure left the Congress weak and gave it no ability to manage federal debt or to maintain a national army. Determined to change that, Madison began studying many different forms of government.

In 1784, Madison was elected to the Virginia House of Delegates and quickly moved to defeat a bill intended to give taxpayer-funded financial support to "teachers of Christian religion." He was successful. Over the following years, Madison spearheaded a movement that pushed for changes to the Articles of Confederation. That effort eventually culminated in the Constitutional Convention of 1787, again in Philadelphia.

At the convention, Madison presented his ideas for an effective government, known as the Virginia Plan. Madison said that the United States needed a strong federal government, which should be split into three branches—legislative, judicial, and executive—and managed with a system of checks and balances, so that no branch could dominate another. During the Constitutional Congress proceedings, Madison took extensive notes and tweaked his plan to make it more acceptable to his fellow delegates. In the end, the Virginia Plan underpinned large parts of the U.S. Constitution.

After the Constitution was written, the document needed to be ratified by 9 of the 13 states. Initially, it was met with resistance, as many states believed that it gave too much power to the federal government. To promote the Constitution's ratification, Madison collaborated with Founding Fathers Alexander Hamilton and John Jay. Together, they wrote a series of anonymous essays supporting the Constitution, titled *The Federalist*—better known today as *The Federalist Papers*.

After the publication of 85 essays and extensive debate in the Constitutional Convention with the anti-federalists (who produced their own writings, later known as *The Anti-Federalist Papers*), the U.S. Constitution was signed in September 1787. The document was eventually ratified in 1788, after New Hampshire became the ninth state to ratify the Constitution. In 1790, the new federal government became functional. The innovative and enlightened ideas of the U.S. Constitution have stood up to the test of time; today, it is the world's oldest written constitution still in operation.

Madison was immediately elected to the newly formed House of Representatives and began working on a draft of the

Bill of Rights, 10 amendments to the Constitution that spelled out the fundamental rights held by every U.S. citizen. They included, among others, freedom of speech and of religion and the right to bear arms. In the Ninth Amendment, Madison also stipulated the existence of unenumerated rights. After substantial debate, Madison's work paid off, and the Bill of Rights was enacted in 1791. These amendments were unique for their time because they stipulated that governments do not grant rights to the citizenry. Instead, it is the citizens who grant powers to governments to protect the people's "pre-existing" rights.

After a disagreement with Federalist leader Alexander Hamilton over Hamilton's proposal to establish a national bank, Jefferson and Madison founded the Democratic-Republican party in 1792. It was the first opposition political party in the United States. Madison left Congress in 1797. He returned to politics in 1801, joining President Thomas Jefferson's cabinet. As secretary of state, Madison oversaw the purchase of the Louisiana Territory from France in 1803, which doubled the size of the new nation.

Between 1809 and 1817, Madison served as the fourth president of the United States. Much of his presidency was marred by overseas problems. In 1812, Madison issued a war proclamation against Great Britain. Trade between the United States and Europe ceased, which severely hurt American merchants. At the same time, New England threatened to secede from the Union. In August 1814, Madison was forced to flee the new capital of Washington after British troops invaded and burned down several buildings, including the White House, the Capitol, and the Library of Congress.

In 1815, the war ended in a stalemate. After two terms as president, Madison returned to Montpelier in 1817 and never left Virginia again. He remained an active and respected writer. In 1826, he became rector (head of the governing board) of the University of Virginia, which was founded by Thomas Jefferson in 1819.

Like many of his contemporaries in the South, Madison owned slaves. However, he worked to abolish the practice of slavery. Under his leadership, the federal government purchased slaves from slaveholders and resettled them in Liberia. Madison spent his last years bedridden and in ill health. In June 1836, he died from heart failure. He was 85 years old.

Madison was instrumental in the drafting of the U.S. Constitution and the Bill of Rights. The U.S. Constitution was the world's first single-document constitution. The Enlightenment principles of individual rights and freedoms that it championed became the basis for dozens of fother liberal constitutions created by governments across the world. The legal framework Madison created has protected countless people from government abuses.

11
William Wilberforce

Statesman and successful abolitionist

William Wilberforce was a leading 18th-century British abolitionist and politician. His efforts helped to ban the slave trade in 1807 and abolish slavery in the British Empire in 1833, freeing millions of formerly enslaved people.

Wilberforce was born on August 24, 1759, in Kingston upon Hull, England. His father was a wealthy merchant. At the age of 17, Wilberforce enrolled at the University of Cambridge. Following the deaths of his grandfather and uncle in 1776 and 1777, respectively, Wilberforce received two large inheritances. Now independently wealthy, he lived a relatively carefree life while studying at Cambridge. He was well-known within the university's social scene and became friends with the future prime minister, William Pitt the Younger.

After graduating in 1780, Wilberforce decided to seek political office, and at the age of 21, he became the member of Parliament for Hull. He was independent of any political party, stating he was a "no party man." In his first four years in Parliament, he admitted he "did nothing to any purpose. My own distinction was my own darling object." As in his university days, Wilberforce was known in many social circles and was fond of drinking and gambling.

In 1785, Wilberforce traveled to Europe with his sister and mother for a vacation. During his time abroad, he read *Rise and Progress of Religion in the Soul*, which had a profound impact on his life. He embraced evangelical Christianity, lost interest in card games and drinking, began to get up early to read the Bible, and decided to commit his future life to work in the service of God.

Thereafter, his political views were guided by his faith and his desire to promote Christian ethics. And so began his lifelong concern with social reform.

In 1786, Wilberforce began to play an active role in the Abolitionist Movement. In 1787, he wrote in his journal that God had set before him the objective of suppressing the slave trade. A group of evangelical abolitionists known as the Clapham Sect soon acknowledged Wilberforce as their leader.

In 1789, he introduced 12 resolutions against the slave trade to the British Parliament's House of Commons. Even though he was often supported by Pitt and the renowned member of Parliament and philosopher Edmund Burke, Wilberforce's measures failed to gain majority support. Wilberforce remained resilient and introduced eight anti-slavery bills between 1791 and 1805. Alas, all were defeated.

After the death of Pitt in 1806, Wilberforce tried once more, but this time, rather than calling for an outright ban on slavery, he strategically pushed a bill that would make it illegal for slaveowners to trade slaves with the French colonies. The bill passed, and this smaller step worked to weaken the power of shipowners in the slave trading business, making it easier for Wilberforce to pass more significant legislation in the future.

In 1807, Wilberforce managed to get the Slave Trade Act passed through both Houses of Parliament. However, the 1807 act banned only slave trading, and many people continued to be held in bondage.

For the remainder of his life, Wilberforce campaigned for the rights of slaves. Despite failing health, he remained integral to the Abolitionist Movement. In 1825, Wilberforce declined a peerage and resigned his seat for health reasons.

On July 26, 1833, the Whig government, under the leadership of Earl Grey, introduced a bill for the abolition of slavery and formally acknowledged Wilberforce in the process. The bill would outlaw slavery in most parts of the British Empire. After hearing the happy news, Wilberforce died just three days later.

Wilberforce's work was integral to the abolition of slavery throughout the British Empire, the leading global power of the day. Thereafter, British ships and the Royal Marines proceeded to extinguish slavery throughout much of the world. The Royal Navy's West African squadron, which was tasked with suppressing the Atlantic slave trade, patrolled the coast of West Africa, boarded slave ships, and blockaded ports. Between 1808 and 1860, the West African squadron captured more than 1,600 slave ships and freed 150,000 enslaved Africans. These efforts helped end the Atlantic slave trade and set in motion the abolition of slavery by governments across the world. Millions were freed from slavery and the idea that every human being ought to be free became widespread. For the first time in human history, millions were freed from bondage and the dignity of every human being was affirmed.

12

Charles Babbage and Ada Lovelace

Early computer designers

Charles Babbage and Ada Lovelace were 19th-century English mathematicians and pioneers of computing. Babbage is often called the father of computing for conceiving the first automatic digital computer. Lovelace, building on Babbage's work, was the first person to recognize that computers could have applications beyond pure calculation. She has been dubbed the first computer programmer for creating the first algorithm for Babbage's machine. Babbage and Lovelace's work laid the groundwork for the modern-day computer industry. Without their contributions, much of the technology we have today would likely not exist.

Charles Babbage was born on December 26, 1791, in London. His father was a successful banker, and Babbage grew up in affluence and attended several of England's top private schools. His father ensured that Babbage had many tutors to assist with the boy's education. As a teenager, Babbage joined the Holmwood Academy in Middlesex. The academy's large library helped him develop a passion for mathematics. In 1810, Babbage began studying mathematics at the University of Cambridge.

Before arriving at Cambridge, Babbage had already learned much of contemporary mathematics and was disappointed by what he considered a low level of mathematics being taught at the university. In 1812, Babbage and several friends created the Analytical Society, which aimed to introduce to England new developments in mathematics that were occurring elsewhere in Europe. He graduated from Cambridge in 1814.

Babbage's reputation as a mathematical genius quickly developed. In 1815, he began lecturing on astronomy at the Royal Institution, an organization focused on scientific education and research in Westminster, London. The following year, he was elected a fellow of the Royal Society. Despite several successful lectures at the Royal Institution, Babbage struggled to find a full-time position at a university. Throughout early adulthood, therefore, he had to rely on financial support from his father. In 1820, Babbage was instrumental in creating the Royal Astronomical Society, which aimed to support and promote astronomical research.

In the early 19th century, mathematical tables—lists of numbers showing the results of calculations—were central to engineering, astronomy, navigation, and scientific research. However, at the time, all calculations in mathematical tables were done by humans and mistakes were commonplace. Given this problem, Babbage wondered if he could create a machine to mechanize the calculation process.

In 1822, in a paper presented to the Royal Astronomical Society, Babbage outlined his idea for creating a machine that could automatically calculate the values needed in astronomical and mathematical tables. The following year, he obtained a government grant to build a machine that could automatically calculate a series of values of up to 20 decimal places. He called it the "difference engine."

In 1828, Babbage became the Lucasian Professor of Mathematics at the University of Cambridge. He was largely inattentive to his teaching responsibilities and spent most of his time writing papers and working on the difference engine. In 1832, Babbage and engineer Joseph Clement produced a small working model of the difference engine. The following year, plans to build a larger, full-scale engine were scrapped when Babbage began to turn his attention to another project.

In the mid-1830s, Babbage began to develop plans for what he called the "analytical engine," which would become the forerunner to the modern digital computer. Whereas the difference engine was designed for mechanized arithmetic (essentially, an early calculator capable of only addition), the analytical engine would be able to perform any arithmetical operation by inputting instructions from punch cards—long rectangular strips of card stock paper containing code in the form of small holes punched in specific locations. The punch cards would be able to deliver instructions to the mechanical calculator as well as store the results of the computer's calculations.

The analytical engine was designed to function as an automatic mechanical, digital computer that was fully controlled by a preset program. Babbage initially envisioned the analytical engine to apply only for pure calculation. That soon changed, thanks to the work of Ada Lovelace.

Augusta Ada King, Countess of Lovelace (née Byron), was born on December 10, 1815, in London, England. She was the only legitimate child of poet and member of the House of Lords Lord Byron and mathematician Anne Isabella Byron. However, just a month after Ada's birth, Byron separated from Lovelace's mother and left England. Eight years later, he died from disease while fighting on the Greek side during the Greek War of Independence.

Throughout her early life, Lovelace's mother raised Ada on a strict regimen of science, logic, and mathematics. Although frequently ill, and bedridden for nearly a year at the age of 14, Lovelace became fascinated by machines. As a child, she would often design fanciful boats and flying machines.

As a teenager, Lovelace honed her mathematical skills and quickly became acquainted with many of the top intellectuals of the day. In 1833, her tutor, Mary Somerville, introduced her to Charles Babbage. The two quickly became friends. Lovelace was fascinated with Babbage's plans for the analytical engine, and Babbage was so impressed with Lovelace's mathematical ability that he once described her as "the enchantress of numbers."

In 1840, Babbage visited the University of Turin to give a seminar on his analytical engine. Luigi Menabrea, an Italian engineer and future prime minister of Italy, attended Babbage's seminar and transcribed it into French. In 1842, Lovelace spent nine months translating Menabrea's transcription of Babbage's lecture into English. She added her own detailed notes, which ended up making the work three times longer than the original article.

Published in 1843, Lovelace's notes described the differences between the analytical engine and previous calculating machines—mainly the former's ability to be programmed to solve any mathematical problem. Lovelace's notes also included a new algorithm for calculating Bernoulli numbers—a sequence of rational numbers that are common in number theory. Because Lovelace's algorithm was the first to be created specifically for use on a computer, she was the world's first computer programmer.

Whereas Babbage designed the analytical engine for purely mathematical purposes, Lovelace was the first person to see a potential use of computers that went far beyond number-crunching. She realized that the numbers within the computer could be used to represent other entities, such as letters or musical notes. Thus, she also prefigured many of the concepts associated with modern computers, including software and subroutines (sequences of instructions within a computer program).

The analytical engine was never built in Babbage's or Lovelace's lifetime. However, the lack of construction was due not to any design flaws, but to funding problems and personality clashes between Babbage and potential funders.

Throughout the remainder of his life, Babbage dabbled in many fields. He twice unsuccessfully ran for a seat in Parliament. He wrote over a dozen books, including one on political economy that explored the commercial advantages of the division of labor. He was fundamental in establishing the United Kingdom's modern postal system. He also invented an early type of speedometer and the locomotive cowcatcher (the metal frame that attached to the front of trains to clear the track of obstacles). On October 18, 1871, Babbage died at his home in London. He was 79 years old.

After translating Babbage's lecture, Lovelace began working on several projects, including one that involved creating a mathematical model for how the brain gives humans the capacity for thought and how it interacts with the rest of the nervous system, although she never achieved that objective. On November 27, 1852, Lovelace died from uterine cancer. She was just 36 years old.

During his lifetime, Babbage declined two noble titles, a knighthood and a baronetcy. In 1824, he received the Gold Medal from the Royal Astronomical Society "for his invention of an engine for calculating mathematical and astronomical tables." Since the deaths of Babbage and Lovelace, many buildings, schools, university departments, and awards have been named in their honor.

Thanks to the work of Babbage and Lovelace, the field of computing was changed forever. Without Babbage's work, the world's first automatic digital computer would not have been conceived when it was. Likewise, many of the main elements that modern computers use today likely would not have been developed until much later. Without Lovelace, it may have taken humanity much longer to realize that computers could be used for more than just mathematical calculations. Together, Babbage and Lovelace laid the groundwork for modern-day computing, which is used by billions of people across the world and underpins much of our progress today.

13
Richard Cobden

Champion of free trade

Richard Cobden was a 19th-century British politician, textile manufacturer, and campaigner for free trade and peace. Thanks to his work repealing the Corn Laws, which placed a tariff on imported food and grain, Cobden helped to turn Britain, the leading global power at the time, into a free-trading nation. This act set in motion an era of increased liberalization of international trade that continues to this day and has helped billions of people rise out of poverty. Cobden also helped to create the first modern free trade agreement, the basis of which is still emulated in trade deals today.

Richard Cobden was born on June 3, 1804, at his family's farmhouse in rural Sussex, England. His parents were relatively poor farmers, which meant that Cobden and his 10 siblings had a modest upbringing. Cobden, despite having a passion for learning, received limited formal education. As a child, he first attended a local, low-cost private school, often referred to as a dame school. At age 14, he became a clerk in his uncle's textile warehouse in London. While in London, Cobden regularly read in local libraries.

In 1828, Cobden and two other young men started a company selling calico prints, which are patterned textiles usually featuring bright colors and repeating patterns, in London. In 1831, the three men opened a calico-printing mill in Lancashire, in the north of England. The new business venture flourished and soon had three establishments: the printing mill in Lancashire and shops in London and Manchester. The newly affluent Cobden moved to Manchester in 1832 and became responsible for the Manchester outlet. It is estimated that while in Manchester, Cobden earned between £8,000 and £10,000 per year (approximately $930,000 to $1.17 million in 2023 U.S. dollars.)

Despite his thriving business, reading and writing absorbed much of Cobden's time. In the early 1830s, Cobden, drawing on works from classical economists such as Adam Smith and David Ricardo, published numerous letters on issues related to economics and free trade in the *Manchester Times.*

In 1833, Cobden began traveling the world. He visited much of Europe, the United States, and the Middle East. While on his travels in 1835, Cobden wrote an influential pamphlet titled *England, Ireland, and America*. In the pamphlet, he advocated for a new approach to foreign policy based on free trade, peace, and noninterventionism.

In 1837, Cobden returned to England and quickly became a popular figure in Manchester's political and intellectual circles. Perhaps because of his lack of educational opportunities as a child, Cobden was passionate about public education as well as free trade. He regularly spoke about the issue in cities across the north of England and campaigned to persuade the authorities to invest more in schools. On one of these campaigns, Cobden met John Bright, a talented orator and future Liberal Party politician.

Before long, Cobden began advocating for the repeal of the Corn Laws. Enacted in 1815, the Corn Laws were tariffs on food and grain imported into Britain. They kept grain prices artificially high to favor domestic producers. Cobden argued that these laws raised the price of food and the overall cost of living for the British public and hampered the growth of other economic sectors besides agriculture. Landowners, mostly members of the Conservative Party, usually supported the Corn Laws. Cobden argued that the tariffs unjustly helped landowners at the expense of the middle and working classes.

In 1838, Cobden joined the Manchester Anti-Corn Law Association, an organization advocating for repealing tariffs. Cobden proved himself a valuable organizer, and in 1839, along with Bright, he played a crucial role in turning the association into a national organization called the Anti-Corn Law League. The organization soon hosted mass meetings, and Cobden, with Bright's support, spoke to audiences across the country. The league also printed a weekly newsletter, published hundreds of books and pamphlets emphasizing the merits of free trade, and submitted petitions to Parliament urging the end of protectionist trade policies.

In 1841, after the Conservative Sir Robert Peel defeated incumbent Whig Prime Minister William Lamb, a general election was called, and Cobden became the member of Parliament (MP) for Stockport. The economic hardship of the recession from 1840 to 1842 pushed more people to oppose the Corn Laws. As an MP, Cobden could directly confront Prime Minister Peel in debate, during which he often blamed Peel and the Conservative Party for the nation's poor economic outlook.

In 1846, after years of supporting the tariffs, Peel reversed his stance and called for the repeal of the Corn Laws. He created a coalition of the Conservative leadership and got about one-third of the party's MPs to vote to abolish the laws. This Conservative coalition, with the backing of the majority of Whig MPs, passed the motion to abolish the Corn Laws in the House of Commons by 98 votes on May 16, 1846. MPs voted voted 327 to 229 leading to a majority of 98—in total there were 656 MPs at the time. The tariffs were abolished soon after the vote.

Peel's decision to abolish the Corn Laws split the Conservative Party and led to the fall of his government. In his resignation speech as prime minister, Peel acknowledged Cobden as the man primarily responsible for changing his mind about the tariffs. Thanks to the repeal of the Corn Laws, millions of Britons, many of whom lived in extreme poverty, finally had access to cheaper foodstuffs from abroad. Moreover, the repeal of the Corn Laws forced many of Britain's colonies to embrace free trade.

After his enormous success in changing public opinion and policy, Cobden toured France, Spain, Italy, Germany, and Russia, speaking at numerous demonstrations and meeting with leading politicians. In 1847, he returned to England and became an MP for West Riding in Yorkshire.

Despite the prestige awarded to Cobden after the Corn Laws' repeal, the seven-year fight to end them left him financially ruined. To help Cobden, in 1847, a public subscription—whereby members of the public voluntarily give someone money—raised £80,000 (approximately $9.3 million in 2023 U.S. dollars) for Cobden. With some of these proceeds, he bought and refurbished the Sussex farmhouse where he was born.

Throughout the early 1850s, Cobden wrote a series of pamphlets about international peace and its connection with free trade. Cobden was an outspoken critic of the Opium War, and in 1852, when Britain declared war on Burma, he proclaimed, "I blush for my country." In 1853, Cobden wrote one of his best-known pamphlets, *1793 and 1853 in Three Letters,* which urged his political contemporaries to learn from past errors and not go to war with France.

In 1859, with tensions between Britain and France high, Michel Chevalier, a French statesman, urged Cobden to persuade the French emperor Napoleon III of the benefits of free trade. With the blessing of the chancellor of the exchequer (the equivalent of the secretary of the treasury in the United States), William Gladstone, Cobden met with Napoleon III to discuss a potential Anglo-French free trade deal.

Cobden worked tirelessly to create this trade deal. In time, Napoleon III was receptive to Cobden's arguments, and on January 23, 1860, Britain and France signed the Cobden-Chevalier Treaty. This trade deal helped to cool tensions between the two great powers and helped prevent another Anglo-French war. The Cobden-Chevalier Treaty also incorporated the "most favored nation" clause, which stated that neither nation could impose any limits on imports or exports that did not also apply to other countries. This clause was copied in many later trade agreements and remains a common stipulation in international trade deals today. Princeton University economist Gene Grossman described the treaty as the "first modern trade agreement."

Cobden was offered many honors for his work. The British government offered him a baronetcy and a seat on the privy council, yet Cobden repeatedly declined. The strain of negotiating the Cobden-Chevalier Treaty had a marked impact on Cobden's health. Cobden died on April 2, 1865, at his apartment in London. Before he passed, Cobden tried desperately to leave his bed to attend Parliament and vote against further spending on national fortifications. The day after he died, the prime minister, Lord Palmerston, stated that Cobden "was an ornament to the House of Commons and an honor to England."

Cobden was buried in West Sussex on April 7, 1865. In the following days, large crowds of mourners, including many prominent politicians, came to his gravesite to pay their respects.

The repeal of the Corn Laws marked a fundamental shift in the British Empire—which then spanned large parts of the globe—toward free trade. That policy alleviated the hunger and suffering of millions of people and set a precedent for free trade treaties around the world to follow. Cobden's influence on the creation of the Cobden-Chevalier Treaty laid the foundation for modern trade agreements that continue to shape and enrich the world.

14

John Stuart Mill

Advocate of classical liberalism

John Stuart Mill was a 19th-century English philosopher, parliamentarian, and political economist. Throughout his life, he advocated for greater freedom of expression, equality between the sexes, and the abolition of slavery. As a member of the British Parliament, Mill presented the House of Commons with the first mass petition in favor of women's suffrage, which helped inspire the creation of numerous suffragette campaigns throughout the world. One of Mill's greatest contributions to philosophy was his *harm principle*, which holds that an action by a person should be legally prohibited only if it causes harm to other individuals. The *Stanford Encyclopedia of Philosophy* describes him as "the most influential English-speaking philosopher of the nineteenth century."

John Stuart Mill was born on May 20, 1806, in London. His father, James Mill, was a close friend of philosopher Jeremy Bentham (see chapter 8). John Stuart Mill had an extraordinary upbringing. His father educated him with the intention of creating an intellectual genius who would lead the next generation of radical and utilitarian thinkers.

At the age of three, the younger Mill began learning ancient Greek. At eight years old, he was learning Latin. By the time he was 12, Mill had read most of the classical canon and began an in-depth study of scholastic logic. John Stuart Mill was deliberately prevented from associating with children his own age except for his siblings. For amusement, he would often read treatises on experimental science.

In 1820, Mill took a year-long trip to France, where he stayed with the family of Samuel Bentham, the brother of Jeremy. Excerpts from the diary he kept show that Mill spent much of his time in France meticulously studying chemistry, math, and the French language. On his return to Britain in 1821, Mill began studying Roman law with renowned English legal theorist John Austin. He also began to study political economy with David Ricardo, one of history's most influential classical economists.

As a nonconformist (a Protestant who did not belong to the Church of England), Mill was not eligible to enroll at the elite universities of Oxford and Cambridge. In 1823, 17-year-old Mill decided to follow in his father's footsteps and began working for the British East India Company.

Mill stayed at the East India Company for more than 35 years, first as an assistant examiner. Then, following the death of his father in 1836, Mill became responsible for the company's relations with the Indian states. Although the work that Mill completed in his day job had little historical significance, it allowed him plenty of time for his personal writings.

After 21 years of friendship, Mill married Harriet Taylor in 1851. She was a philosopher and women's rights advocate. In 1858, the East India Company was dissolved. The newly unemployed Mill moved to Avignon, France, where he continued to write full-time. In 1859, he published one of his best-known works, *On Liberty.* He dedicated the book to Taylor, who had died the year before. He acknowledged her as a huge influence on his thinking, especially on women's rights.

On Liberty focuses on the nature and limits of the power that governments may rightfully exercise over an individual. The extent of a government's power, Mill argued, should be based on the harm principle. In *On Liberty*, Mill also claims that freedom of speech is a necessary condition for a society to make intellectual and social progress. He believed that society can never be sure that a banned opinion does not contain at least some elements of truth. Therefore, people should be free to state any opinion they wish. Mill declared that even if an opinion is false, individuals are more likely to abandon their incorrect views through open discourse. He noted that the truth can be better understood and prevented from becoming mere dogma if individuals continuously reexamine their beliefs. The book was an enormous success, and Mill soon became a well-known public intellectual.

In 1861, Mill completed the essay "The Subjection of Women." It advocated for complete equality of the sexes. Mill believed that the oppression of women was a relic from ancient times and "one of the chief hindrances to human improvement." The essay made Mill one of the earliest male proponents of gender equality. Mill also expressed his opposition to slavery and his support for its abolition in the United States.

In the same year, Mill also published *Considerations on Representative Government,* in which he stood up for proportional representation, the single transferable vote, and the extension of suffrage to women.

In 1863, Mill published *Utilitarianism*. The book is a strong defense of utilitarian ethics, a philosophy that, according to Mill, suggests "that actions are right in the proportion as they tend to promote happiness, wrong as they tend to produce the reverse of happiness." Like Bentham, Mill contended that there should be legislation favoring animal welfare and that the economic system of free markets was preferable to a planned economy.

In 1865, the Liberal Party asked Mill to become its candidate for Parliament for Westminster (the area of London that is home to the Houses of Parliament, which was its own constituency until 1918). He agreed on the condition that he would not canvass or contribute financially to his own campaign. He also stated that if he were elected, he would not speak for local interests but would instead use his position to "serve as the conscience of his society." He said he would continue to support women's suffrage.

Despite his lack of campaigning, Mill won the election and used his time in Parliament to advocate for land reform in Ireland and women's rights for all. In 1866, he presented to Parliament a petition in favor of female enfranchisement with more than 1,500 signatures that had been collected by the Women's Suffrage Committee. During this time, Mill also campaigned for universal education.

In 1866, the Second Reform Act—a bill designed to expand the electorate by loosening the property qualifications—was debated in Parliament. Mill used the Second Reform Act as an opportunity to introduce equal voting rights for men and women by proposing an amendment that replaced the instances of "man" with "person," a change that would have enfranchised some property-owning women.

Unfortunately, the amendment was defeated, but Mill's advocacy prompted widespread debate surrounding women's suffrage and inspired the creation of several political campaigns for female enfranchisement. Mill later described the amendment as "perhaps the only really important public service I performed in the capacity as a Member of Parliament."

In the general election of 1868, Mill lost reelection and returned to France to study and write. On May 8, 1873, Mill died of erysipelas, a serious skin infection, in Avignon and was buried alongside his wife.

John Stuart Mill is remembered as one of the most important and influential philosophers of the 19th century. His enormous body of work continues to shape political thought and discourse. Mill's advocacy for women's rights, the harm principle, and freedom of speech has helped to create less oppressive and more egalitarian laws in many nations.

15

John Snow

Father of epidemiology

John Snow was a 19th-century English physician who is considered by many to be the father of epidemiology and who made significant breakthroughs in anesthesiology. Following a series of cholera outbreaks in London, Snow was the first person to use maps and data records to track the spread of a disease back to its source. His work provided a foundation for the science of epidemiology, and he improved the way humanity confronts public health emergencies.

Snow was born on March 15, 1813, in York, England, and was the oldest of eight siblings. His father was a coal yard laborer, and Snow was raised in one of the city's poorest neighborhoods. As a child, he was exceptionally bright, with a strong aptitude for mathematics. His mother, recognizing Snow's academic talents, used a small amount of money she inherited to send him to a nearby private school.

Snow excelled at school. In 1827, at the age of 14, he attained an apprenticeship under Dr. William Hardcastle in Newcastle, about 100 miles from York. In 1831, a cholera epidemic began to spread across Europe. By 1832, Killingworth, a nearby village, was severely afflicted with the disease. With Hardcastle overwhelmed by patients, Snow was sent to the village to attempt to treat the victims.

Cholera causes its victims to suffer from severe diarrhea and vomiting, which leads to rapid dehydration. It can be fatal within just a few hours. Unfortunately, there was little that Snow could do to help the cholera-stricken villagers. The typical medical treatments of the day, which included laxatives, opium, brandy, and peppermint, were all ineffective. A few months later, the epidemic ended. In total, the outbreak killed more than 50,000 Britons. The early experience of helplessness in combating cholera made a significant impact on Snow.

In 1832, Snow began to work as an assistant to a colliery surgeon in County Durham. In 1836, he enrolled in the Hunterian School of Medicine in London. A year earlier, Snow had signed an alcohol abstinence pledge and became a supporter of teetotalism. He also became a vegetarian and would drink only water that had first been boiled so it was "pure."

In 1837, Snow started working at Westminster Hospital and a year later was admitted as a member of the Royal College of Surgeons of England. In 1844, he received a doctorate in medicine from the University of London. After graduation, Snow began to work as a surgeon and general practitioner.

For several years, Snow meticulously studied the effects of different anesthetics. In his day, it was typical for a surgeon to use either too little anesthetic, which could cause the patient to wake up during surgery, or too much, which could cause death. Snow was one of the first physicians to study and calculate the dosages of ether and chloroform needed for surgery. For many years, he tested the effects of ether and chloroform on himself. By taking notes on how long he was unconscious after different dosages, he was eventually able to work out the optimal amount of anesthetic that patients could handle.

After creating an anesthetic inhaler and publishing his findings in 1847 in a textbook titled *On the Inhalation of the Vapour Ether in Surgical Operations*, Snow quickly gained fame as Britain's most accomplished anesthesiologist. Snow's fame eventually led him to administer chloroform to Queen Victoria during the birth of her last two children, Prince Leopold and Princess Beatrice.

Despite Snow's impressive accomplishments in the field of anesthesiology, his most important work came a few years later, following a series of cholera outbreaks in London. In the mid-19th century, most physicians thought that diseases such as cholera or the plague were caused by "miasmas," or air pollution. The germ theory of disease had yet to be developed, but Snow theorized that diseases were likely caused by invisible tiny parasites.

In 1848, when a new outbreak of cholera struck London, Snow decided to track the disease to its source to find out how it spread. After examining several patients, he realized that their first symptoms were nearly always digestive problems. He theorized that the disease must have been ingested through food or water. Had the disease been spread from polluted air, as the supporters of the miasmatic theory believed, then logically, the first symptoms should appear in the nose or lungs.

Moreover, Snow reasoned that the severe diarrhea that was caused by cholera might be the mechanism by which the germs spread. Put differently, if dangerous germs were present in the diarrhea and the diarrhea contaminated the water supply, the germs could then spread to countless new victims. In 1849, Snow decided to publish, at his own expense, a pamphlet that outlined his thoughts on how cholera was spread. It was titled *On the Mode of Communication of Cholera.* Although Snow's work had little effect on his colleagues' thinking, he pushed on with his research regardless.

In August 1854, another cholera outbreak occurred in the Soho neighborhood of London. Snow found that of 73 cholera victims, 61 had drunk water from a single pump located on Broad Street. Snow's microscopic examination of the water from the Broad Street pump proved inconclusive. Undeterred, he plotted the numbers and locations of cholera cases on maps of the area to highlight the correlation between cholera infections and use of the Broad Street pump. The following month, Snow showed his evidence to the authorities and recommended that they remove the pump's handle so that no more water could be drawn from the infected source. Although the authorities were not convinced by Snow's argument, they obliged. Subsequently, the local cholera outbreak quickly ended.

Researchers would later discover that the Broad Street well had been dug just three feet away from an old cesspit, which had begun to leak fecal bacteria—a discovery that helped lend credence to Snow's theory.

In September and October 1854, there was another cholera outbreak in London. Snow swiftly began another project that he called his "Grand Experiment." He began to compare cholera death rates for households whose water was supplied by two different companies: the Southwark and Vauxhall Waterworks Company and the Lambeth Company.

Snow found that the Southwark and Vauxhall Waterworks Company relied on water from sewage-polluted sections of the river Thames. In contrast, the Lambeth Company used water from inlets in the upper Thames, miles from urban pollution. Snow created dot maps and used statistics to highlight the correlation between the quality of the water supplied to different households and incidences of cholera.

Snow enlarged his original 1848 pamphlet into a book that included intricate details of all his studies. In 1855, he published the second edition of *On the Mode of Communication of Cholera.* Today, Snow's studies are considered major events in the history of public health, the first time that maps and data were used to accurately track the spread of a disease back to its source. As such, many consider Snow's 1855 book the foundation of epidemiology. However, despite the work's significance, Snow's critics and public health officials at the time remained unimpressed, arguing that the enormous quantity of water in the Thames was large enough to dilute any waterborne poison.

Snow's foresight wasn't truly appreciated until the 1860s, when pioneering French scientist Louis Pasteur (see chapter 17) successfully proved the germ theory of disease. Unfortunately, Snow never got to see his work become widely accepted, because he died from a stroke on June 16, 1858. He was just 45 years old.

John Snow was one of the great physicians of the 19th century. During his short life, he wrote more than 100 books, pamphlets, and essays on a variety of medical topics. He is widely considered the founder of epidemiology. Snow's methods have been copied all over the world and have been used to quell, or at least slow down, many potentially catastrophic outbreaks of deadly diseases, likely saving millions of lives.

16

Frederick Douglass

A leading U.S. abolitionist

Frederick Douglass was an abolitionist and social reformer who is widely considered one of the foremost human rights leaders of the 19th century. As a former slave who became a consultant to President Abraham Lincoln and later to President Andrew Johnson, Douglass helped to convince both presidents of the necessity of equal rights for black Americans.

Douglass's relentless advocacy for equality under the law helped to shift public opinion against slavery in the United States, and his influence in the creation and ratification of the Reconstruction Amendments—a series of constitutional amendments enacted after the Civil War that ensured equal freedom and voting rights for black Americans. While these amendments were not widely followed and a period of formal state oppression continued, their enactment lay the foundation to a better, more prosperous future for millions of people.

Frederick Augustus Washington Bailey (later changed to Douglass) was born enslaved around 1818 in Talbot County, Maryland. As was common with slaves, the exact year and date of Douglass's birth are unknown. He later chose to celebrate his birthday on February 14, based on his memory of his enslaved mother, who called him her "little Valentine." Douglass was separated from his mother by the slave master when he was an infant and saw her only a few times before her death in 1825.

Until the age of six, Douglass was raised by his grandmother, who was also enslaved. In 1824, Douglass was moved to a plantation where Aaron Anthony, a white man who Douglass suspected might be his father, worked as an overseer. Throughout his childhood, Douglass was transferred by various slaveowners to many other plantations. After Anthony's death in 1826, Douglass was given to Thomas Auld, who sent him to serve his brother, Hugh Auld, in Baltimore.

When Douglass was 12, Hugh Auld's wife, Sophia, began teaching Douglass the alphabet. Sophia made sure that Douglass was properly fed and clothed and slept in a comfortable bed. However, over time, Hugh Auld convinced his wife that slaves should not be educated. Education, he thought, would encourage slaves to desire freedom. Before long, Douglass's lessons stopped.

Despite that setback, Douglass continued to secretly teach himself how to read and write. He found a variety of newspapers, pamphlets, books, and political materials that helped him with his education and development of personal views on freedom and human rights. A couple of years later, Douglass was moved to another plantation. He began hosting a weekly Sunday school, where he taught basic literacy to other slaves. The plantation owner did not mind Douglass's Sunday school, but other plantation owners became increasingly agitated at the idea of slaves being educated. One Sunday, armed with clubs, the neighboring plantation owners burst in on the gathering and disbanded it.

On January 1, 1833, Auld sent Douglass to Edward Covey, a farmer with a reputation for harshly beating slaves. Covey often abused Douglass but after the beatings became more intense, Douglass, just 16 at the time, fought back. Douglass won and Covey never attempted to beat him again. Douglass saw the fight as a life-changing event, and he introduced the tale in his autobiography thusly: "You have seen how a man was made a slave; you shall see how a slave was made a man."

On Christmas Day, 1833, Douglass's term of service with Covey ended, and on January 1, 1834, he began laboring for slave-owner William Freeland. After Douglass attempted to escape from slavery in 1835, the Auld family sent him to work at a shipyard in Baltimore. He tried again to flee from slavery in 1833 and 1836. In 1837, Douglass met and fell in love with Anna Murray, a free black woman living in Baltimore. On September 3, 1838, Murray gave Douglass a sailor's uniform, train tickets, and identification papers from a free black seaman. Douglass successfully got away from Maryland by taking a train and a steamboat north to New York City. After 11 days in New York, Douglass and Murray married and resettled in New Bedford, Massachusetts, which had a large free black community. Initially, the couple adopted the name Johnson. Later, they changed it to Douglass in an effort to elude slave hunters who were after Frederick.

In 1839, Douglass became a licensed preacher, which helped him hone his public speaking skills. He also started to attend meetings and protests organized by abolitionist groups. In 1841, Douglass was invited to speak at an abolitionist convention in Nantucket, Massachusetts. The speech went so well that he was invited to work as an agent and speaker for the Massachusetts Anti-Slavery Society.

In 1843, Douglass went on a six-month speaking tour with the American Anti-Slavery Society across the northeastern and midwestern United States. During the tour, the group was often met with violent protests. Many skeptics doubted that Douglass's story was real, arguing that a former slave could not be such an articulate orator.

That criticism led Douglass to write an autobiography in 1845, titled *Narrative of the Life of Frederick Douglass, an American Slave.* The book became an immediate bestseller and was quickly translated and published across Europe. In his lifetime, Douglass published three versions of his autobiography.

Many of Douglass's friends feared that his newfound fame would attract attention from slave hunters acting on behalf of his former master, Thomas Auld. To escape potential problems, Douglass left the United States on a two-year speaking tour of the United Kingdom of Great Britain and Ireland on August 16, 1845.

During his time overseas, Douglass was amazed by the lack of racial discrimination. He later noted that he found himself "regarded and treated at every turn with the kindness and deference paid to white people" and that he was seen not "as a color, but as a man."

Douglass's speaking tour was enormously successful. Many of his new friends and supporters encouraged him to stay in England. However, with his wife still in Massachusetts and more than three million black Americans still in bondage, Douglass was adamant about returning home. Before he left the British Isles, his supporters raised enough money for him to buy his freedom.

In 1847, Douglass returned to the United States and immediately bought his freedom. He also began his own anti-slavery newspaper titled the *North Star.* In 1848, Douglass became involved in the early feminist movement and was the only African American to attend the Seneca Falls Convention, the first women's rights gathering in the United States. Douglass also became an early advocate for school desegregation, asserting that the facilities for black children were vastly inferior to those for whites.

During the U.S. Civil War (1861–1865), Douglass became an adviser to President Abraham Lincoln. He convinced Lincoln that, because the goal of the Civil War was to end slavery, African Americans ought to be able to fight in the war. Douglass also urged the president to issue the Emancipation Proclamation, which declared that slaves in Confederate territory would be free if they escaped to Union-controlled territory. Douglass further advised President Lincoln, and later President Johnson, to push for a series of amendments to the U.S. Constitution that would protect the rights of black Americans.

In 1865, the Thirteenth Amendment, which outlawed slavery, was ratified. In 1868, the Fourteenth Amendment provided citizenship and equal rights to black Americans. In 1870, the Fifteenth Amendment prohibited federal and state governments from denying a citizen the right to vote based on race. Together, these three amendments are called the Reconstruction Amendments.

For the rest of his life, Douglass remained dedicated to equality for all. In 1870, he started his last newspaper, the *New National Era.* In 1872, he became the first black American to be nominated for vice president of the United States on the Equal Rights Party ticket. In 1874, he was appointed as the U.S. Marshall for the District of Columbia. Throughout the 1880s, Douglass traveled the world and spoke on human rights and racial equality. At the 1888 Republican National Convention, he became the first African American to receive a vote for president of the United States. From 1889 to 1891, Douglass served as the American consul general to Haiti.

On February 20, 1895, Douglass died of a heart attack. He was 77 years old. Posthumously, he has received dozens of awards and honors. To this day, many schools, parks, scholarships, and statues are named or erected in his honor.

Douglass's work helped shift U.S. public opinion against slavery, and his influence helped to bring about the ratification of the Reconstruction Amendments, which for the first time provided rights to millions of African Americans.

17

Louis Pasteur

Father of microbiology

Louis Pasteur was a 19th-century French scientist who is commonly called the father of microbiology. He is renowned for developing the germ theory of disease; creating the process of pasteurization, which prevents the spoiling of many food products; and changing the way that scientists create vaccines.

Louis Pasteur was born to a poor family in Jura, France, on December 27, 1822. In 1839, he enrolled at the Royal College of Besançon, the same city in which he had attended secondary school. Within a year, Pasteur had earned a bachelor of letters—a degree given to an undergraduate specializing in a field of personal interest without formal teaching. In 1842, he graduated with a degree in science. A year later, he started studying at the *École Normale Supérieure,* a graduate school in Paris. In 1848, Pasteur was appointed professor of chemistry at the University of Strasbourg.

In Strasbourg, Pasteur met his future wife, Marie Laurent. The two married in 1849 and had five children. However, only two of those children survived to adulthood, while the rest died of typhoid. The death of his three children might have motivated Pasteur to study infections and vaccinations.

In 1856, when Pasteur was the dean of the faculty of sciences at the University of Lille, he started to study fermentation to help a local wine manufacturer overcome the problem of alcohol souring. Before Pasteur, people believed in a theory known as "spontaneous generation," which held that life—such as mold—appeared spontaneously from nonliving matter. That faulty reasoning was used to explain why food spoiled and how infections developed.

However, Pasteur's numerous experiments on fermentation went against the prevailing wisdom of spontaneous generation. In one experiment, Pasteur found that grapes that were removed from their skin and wrapped in sterilized cloth never fermented. In another experiment, he allowed air to enter a flask of boiled broth through a long, winding glass tube that made dust particles stick to it. He found that no organisms grew in the broth unless the flasks were tilted and thus exposed to the contaminated walls. Furthermore, Pasteur found that heating of beverages to a temperature ranging from 140 degrees Fahrenheit to 212 degrees Fahrenheit (60 degrees Celsius to 100 degrees Celsius) killed the bacteria in those liquids. His first successful test was completed on April 20, 1862, and the process he developed came to be known as pasteurization. He patented his discovery in 1865.

Pasteur then turned his attention to the development of vaccines. He was the first scientist to use artificially weakened viruses as vaccines. Pasteur and his colleagues injected chickens with cultured cholera microbes. After many experiments, the team discovered that if the birds were injected with live cholera microbes after they had already been injected with a weaker strain of cholera, the chickens would remain healthy. Pasteur then went on to develop a vaccine for anthrax in 1881. In 1885, he successfully developed a vaccine for rabies.

By 1888, Pasteur had received enough donations to open the Pasteur Institute, a private foundation dedicated to the study of biology, microorganisms, diseases, and vaccines. He remained director of his institute until his death on September 28, 1895.

Pasteur became a grand officer of the Legion of Honor in 1878, France's highest order of merit. He received dozens of honorary awards, and today there are some 30 institutes and several hospitals, schools, and streets named after him around the world. When he died, Pasteur was given a state funeral in the Cathedral of Notre Dame, and his body was interred in a vault beneath his institute, where it still lies.

The work of Louis Pasteur fundamentally changed the world we live in. The proof he provided for the germ theory of disease revolutionized the way we think about human health.

Pasteurization enabled us to preserve beverages and canned foods far longer than was previously thought possible. And finally and crucially, Pasteur revolutionized the development of vaccines.

Much of modern science rests on Pasteur's work. Without him, it is likely that hundreds of millions, if not billions, of people would not be alive today.

18

Joseph Lister

Father of modern surgery

Joseph Lister was a 19th-century British surgeon who is commonly dubbed the father of modern surgery. He introduced new methods of sterilization to hospitals and new medical equipment that transformed hospitals from dangerous places where the risk of further illness was high into places of genuine healing. In hospitals that followed his suggestions, infection rates fell drastically. Since then, Lister's ideas have been adopted as surgical norms around the world.

Joseph Lister was born on April 5, 1827, in Essex, England, to a prosperous Quaker family. His father was a successful wine merchant and amateur scientist who pioneered achromatic (non-color-distorting) lenses that are used in microscopes today.

Lister attended University College London, one of the very few schools in Britain that accepted Quakers at the time, to study botany. He graduated with a bachelor of arts in 1847.

Upon finishing his undergraduate degree, Lister immediately enrolled as a medical student at the same institution. He graduated with honors in 1852 and became a house surgeon at University College Hospital. A year later, he moved to Edinburgh to work as an assistant to James Syme, a renowned surgical teacher at the time. In 1859, Lister assumed the prestigious position of regius professor (a professor who has had royal patronage or appointment) of surgery at Glasgow University.

Ignorance about the spread of disease was rife in the mid-19th century. For instance, it was widely believed that bad air was responsible for infections in patients' wounds. Hospitals would air out their wards at midday to reduce the spread of "miasma"—poisonous vapor filled with particles from decomposed matter that supposedly caused infections. Surgeons did not wash their hands and clothes, boasting about the stains on their dirty operating gowns and referring to the "good old surgical stink."

While working at the University of Glasgow, Lister read a paper by French chemist Louis Pasteur (see chapter 17), who showed that food spoilage may occur under anaerobic (airless) conditions if microorganisms are present. Pasteur debunked the notion that bad air caused infections and recommended filtration, exposure to heat, or chemical solutions to arrest the spread of harmful bacteria.

Because Pasteur's first two suggestions would damage human tissue, Lister sprayed patients' wounds, surgical instruments, and dressings with carbolic acid (C_6H_5OH). He soon noticed that C_6H_5OH dramatically reduced infection rates among his patients. He published his findings in the scientific journal *The Lancet* over a series of six articles in early 1867.

Lister ordered all surgeons under his direction to wear clean gloves, wash their hands and instruments before and after operations, and spray the C_6H_5OH solution in the operating room. His methods quickly caught on across the world. As hospitals and operating rooms became cleaner and wounds were sterilized, rates of infections fell, and millions of lives were saved.

In 1869, Lister returned to Edinburgh to succeed Syme and continued to develop his methods of sterilization. Lister's most famous patient was Queen Victoria, who in 1871 called on Lister to attend to an orange-sized abscess in her armpit. Armed with carbolic acid and a small incision tool, he treated the queen's wound. Later Lister would often joke to his students, "Gentlemen, I am the only man who has ever stuck a knife into the queen!"

Lister continued teaching, researching, and treating patients until his wife passed away in 1893. After that, he said that studying and writing had lost their appeal for him. Lister died on February 10, 1912, at his country home in Kent, at the age of 84.

Lister was decorated with numerous awards and honorary degrees throughout his life. Most notably, he was made president of the Royal Society in 1894 and a baron by Queen Victoria in 1897. Lister's sterilization methods transformed modern surgery and saved untold millions of lives.

19

Wilhelm Röntgen

Discoverer of x-rays

Wilhelm Röntgen was a 19th-century German scientist and the first person to identify electromagnetic radiation in a wavelength that we today know as an x-ray. Today, x-ray machines are common at most medical facilities. They are used to monitor dozens of medical conditions, most commonly to detect fractured bones, heart problems, breast cancer, scoliosis, and tumors. The ability to accurately monitor the internal conditions of our bodies leads to better medical decisions. Every year, x-ray machines are used to help save the lives of millions of people.

Wilhelm Röntgen was born on March 26, 1845, in Lennep, Prussia, in what is now northern Germany. In 1862, he attended a boarding school in Utrecht, the Netherlands. He was expelled in 1865 after he was accused of drawing a disparaging caricature of one of his teachers. Without a high school diploma, he could enroll at a university only as a visitor, not an actual student. The Federal Polytechnic Institute in Zurich, Switzerland, did not require a high school diploma, so, having passed the entrance exams, Röntgen enrolled there as a student of mechanical engineering.

In 1869, Röntgen obtained a PhD and became an assistant to Professor August Kundt, whom he followed first to the University of Würzburg and then to the University of Strasbourg. In 1874, Röntgen qualified to become a lecturer at Strasbourg. He became a full professor in 1876. In 1879, he was named chair (a titled bestowed on a professor who has significantly contributed to the research in a given field) of physics at the University of Giessen. Röntgen moved once again in 1888, to become chair of physics at the University of Würzburg. It was during his time at Würzburg that he made his world-changing discovery.

On November 8, 1895, Röntgen was conducting experiments using a cathode ray tube, a specialized vacuum tube that gives off fluorescent light when an electrical charge passes through it (it was later used in televisions sets). Röntgen noticed that when he used the cathode ray tube, a board on the other side of his lab that was covered in phosphorus began to glow. Intrigued, he covered the tube in a thick black cardboard box to obscure the light that the tube emitted. Röntgen noticed that even after the tube's light had been covered, the phosphorus board continued to glow. It soon became clear that he had discovered a new type of ray. Given the ray's unknown nature, he named it "x-ray" (in mathematics, *x* is often assigned to something unknown).

Röntgen spent the following weeks sleeping and eating in his laboratory as he investigated the properties of these new rays. After numerous experiments, he found that many materials were transparent or translucent when interposed in the path of the rays. These materials included paper, wood, aluminum, and, most importantly for the medical industry, skin and organs. Röntgen used a photographic plate to detail the transparencies of various objects. Two weeks after his discovery, he took the first x-ray picture: a radiograph of his wife's hand. When his wife saw the skeletal image, she exclaimed, "I have seen my own death!"

On December 28, 1895, Röntgen published a paper titled "On a New Kind of Rays" detailing his discovery. Soon after, images of the first x-ray and accounts of Röntgen's work appeared in newspapers globally. Over the next two years, Röntgen published three papers about his experiments. He believed that his discovery should be publicly available and never sought a patent for x-rays. In 1900, at the special request of the Bavarian government, he moved to the University of Munich to chair its physics department.

Röntgen was showered with numerous prizes, medals, and honorary doctorates. In 1901, he was awarded the first Nobel Prize in Physics. After receiving the money given to Nobel Prize winners, he donated all of it to research at the University of Würzburg. On February 10, 1923, Röntgen died from carcinoma of the intestine. He was 77 years old. In 2004, chemical element 111 was named roentgenium in his honor.

Röntgen's discovery of the x-ray fundamentally changed medical practices forever. Every day, his work is being used to help save the lives of people around the world.

20

Kate Sheppard

First successful suffragette leader

Kate Sheppard was the world's first successful suffragette leader. Her tireless work and petitioning of the New Zealand Parliament in the latter half of the 19th century are largely credited for the nation becoming, in 1893, the first country to grant women the right to vote. After New Zealand embraced universal suffrage, Sheppard inspired successful suffrage movements around the world. Today, women almost everywhere have a vote.

Kate Sheppard, née Catherine Wilson Malcolm, was born on March 10, 1847, in Liverpool, England. After her father died in 1862, she moved to Nairn to live with her uncle, a minister of the Free Church of Scotland. Sheppard's uncle taught her the values of Christian socialism that would stay with her the rest of her life. Although the precise details of Sheppard's education aren't known, she possessed an extensive knowledge of both science and law.

In the late 1860s, Sheppard, accompanied by her mother and sister, moved to Christchurch, New Zealand. She quickly became part of the city's intellectual scene and befriended Alfred Saunders, a politician and prominent temperance activist who helped to influence her ideas on women's suffrage. Sheppard married Walter Allen Sheppard, a shop owner, in 1871.

Sheppard was an active member of various religious organizations. She taught at a Sunday school and was elected secretary of the Trinity Ladies' Association, an organization established to visit members of the parish who did not regularly attend church services, in 1884. In 1885, Sheppard became involved in establishing a Christchurch branch of the international Women's Christian Temperance Union (WCTU).

Sheppard's political activism stemmed largely from her interest in temperance. In the late 1880s, she began drafting and promoting petitions to New Zealand's Parliament that would prevent women from being employed as barmaids. After Parliament rejected Sheppard's barmaid petition, she came to believe that politicians would continue to reject petitions put forward by women as long as women did not have the right to vote.

In 1887, Sheppard was appointed national superintendent for franchise and legislation for New Zealand's WCTU, which advocated for prohibition of the sale and distribution of alcoholic beverages (a policy that, as the American experience shows, creates more problems than it purports to solve). By 1888, she was the president of the Christchurch branch of the WCTU. Sheppard quickly became a prominent figure of the women's suffrage movement, and she proved herself a powerful speaker and organizer by hosting political events across New Zealand.

In both 1887 and 1890, there were failed attempts by politicians sympathetic to Sheppard's cause to introduce legislation that would grant women the right to vote. In 1888, she wrote a pamphlet titled *Ten Reasons Why the Women of New Zealand Should Vote*, which the WCTU sent to every member of the House of Representatives. She also wrote pamphlets that were sent to suffrage movements all over the world.

In 1891, Sheppard started making parliamentary petitions to persuade politicians to support giving women the vote. In the same year, she started a petition that contained 10,085 signatures. Sir John Hall, a member of the House of Representatives and a supporter of Sheppard, presented the petition to Parliament alongside a proposed amendment to the existing Electoral Bill that would allow women the right to vote. This amendment passed in the House of Representatives but was rejected in the Upper House. In 1892, Sheppard created another petition with 20,274 signatures, but the women's suffrage amendment once more failed in the Upper House. (The Upper House, or Legislative Council, was dissolved in 1950, and the House of Representatives now serves as a unicameral legislature.)

Eventually, with a petition of 31,872 signatures—the largest petition that the New Zealand Parliament had ever received up to that time—the Electoral Bill of 1893 passed. The enfranchisement of women was signed into law by Governor David Boyle on September 19, 1893.

Sheppard was widely credited for the Electoral Bill of 1893. Seeing the success of the suffrage movement in New Zealand, women's suffrage groups in other countries were inspired to follow in her footsteps. Sheppard sent her writings to suffragettes all over the world. As editor of the WCTU's monthly journal *The White Ribbon,* she promoted suffrage groups abroad. Sheppard was in enormous demand on the speaking circuit. Before going to England in 1903, without a fixed return date, to assist in the suffrage movement there, she gave speeches in Canada and the United States.

Because of failing health, Sheppard moved back to New Zealand in 1904 but embarked on a tour of India and Europe a few years later. In 1916, she was the first person to sign a petition urging New Zealand Prime Minister Sir Joseph Ward to support enfranchisement of women in Britain. Suffrage movements throughout the world copied Sheppard's tactics with enormous success. Australia granted women the right to vote in 1902; Finland did the same in 1906; Norway in 1913; Denmark in 1915; and Austria, Britain, Germany, Poland, and Russia in 1918. The United States finally followed suit in 1920.

The trend continued long after Sheppard's lifetime. Switzerland granted women the vote in 1971, with one canton holding out until 1991. Saudi Arabia allowed women to vote in 2015.

Sheppard died in Christchurch on July 13, 1934. She was 86 years old. Today, her profile is featured on the New Zealand $10 note, and she continues to be considered the world's first successful suffragette leader. Without her tireless work, it is likely that billions of women around the world would have had to wait longer to achieve the political rights they now enjoy.

21

Ronald Ross

Discoverer of how malaria is transmitted

Sir Ronald Ross was a 19th-century British-Indian doctor and polymath who was the first person to prove that malaria is spread by mosquitoes. Ross's groundbreaking discovery laid the foundation for new methods of combating the disease, saving millions of lives.

Ronald Ross was born on May 13, 1857, in Almora, British India, the eldest of 10 children. His father, a general in the British Indian Army, sent Ronald to live on the Isle of Wight with his aunt and uncle when he was eight. As a teenager, he attended a boarding school in Southampton. He developed a passion for writing, mathematics, music, and poetry. Growing up, he won several prizes in these subjects.

Despite Ronald's dreams of becoming a writer, his father insisted that he should become a doctor and sent him to study at Saint Bartholomew's Hospital Medical College in London in 1875. Ross graduated in 1879. After a brief stint as a surgeon on a transatlantic steamship, he joined the Indian Medical Service as a surgeon in 1881. Ross served in the Third Anglo-Burmese War (1885) and, in the following years, was posted all over what are now India and Pakistan.

While on leave in 1888, Ross spent a year studying bacteriology in London. In 1894, during another period of leave in London, Ross met Sir Patrick Manson (commonly known as the father of tropical medicine). Manson introduced Ross to malaria research. Initially, Ross doubted the existence of the malaria parasite, but Manson converted Ross to Laveranity, the belief that malaria parasites become present in an infected person's bloodstream. Manson soon became Ross's mentor, and Ross grew enthusiastic about studying the disease and investigating the hypothesis that mosquitoes are connected with the spread of the disease.

Following Manson's advice that India was the best place to study malaria, in 1895 Ross returned to the subcontinent. Before his luggage cleared the customs office, Ross was in the Bombay Civil Hospital looking for malaria patients to study. In India, Ross bred mosquitoes from larvae, allowed them to feed on patients suffering from malaria, and then dissected the mosquitoes to look for parasites like those found in the patients' blood. However, Ross did not know which of the hundreds of mosquito species in India to study or where the parasite would appear in the insect.

Ross's experiments were interrupted when he was deployed to Bangalore to investigate an outbreak of cholera. In June 1897, he was transferred to Secunderabad to continue malaria research. Then, on August 20, 1897—just two months after his return to Secunderabad and after two years of failure and about 1,000 mosquito dissections—Ross found a malarial parasite inside the gut of an anopheline mosquito. The next day, he confirmed the growth of the parasite in the mosquito. To mark the occasion, on the evening of August 22, he composed a poem to his wife, Rosa, observing, "I find thy cunning seeds, O million-murdering death, I know this little thing, A myriad men will save, O Death, where is thy sting? Thy victory, O Grave?"

Just a week after his discovery, on August 27, 1897, Ross published his initial findings in the *Indian Medical Gazette*. He published more details of his discovery in the December 1897 issue of the *British Medical Journal*.

In 1898, Ross found that the malarial parasite was stored inside the mosquito's salivary gland and released through biting. He also demonstrated the transmission of malaria from infected birds, via the mosquito, to healthy birds—thus establishing the completed life cycle of the malarial parasite. Ross concluded that, as with birds, malaria is transmitted from mosquitoes to humans. Soon after that, Italian zoologist Giovanni Battista Grassi identified *Anopheles* as the genus of mosquito responsible for spreading the disease. Once scientists understood the role of mosquitoes, they were able to develop programs aimed at halting the spread of the illness.

Ross left the Indian Medical Service in 1899 and became a professor at the Liverpool School of Tropical Medicine, where he continued to work on the prevention of malaria in various parts of the world. When Ross won the Nobel Prize in Physiology or Medicine in 1902, he was the first Nobel laureate born outside Europe. He also received a knighthood from King George V in 1911.

In 1912, following a dispute with his managers in Liverpool, Ross moved to London to set up his own private practice. In 1914, following the outbreak of World War I, he was called up to work as a consultant physician for Indian troops stationed in England. For most of the war, he worked for the War Office on the treatment of soldiers with malaria. He was also briefly sent to Egypt to investigate a dysentery outbreak.

After the war, he returned to London, wrote his autobiography, and in 1926 opened the Ross Institute and Hospital for Tropical Diseases, where he was director-in-chief. He continued traveling across Asia until he had a stroke in 1927. Ross passed away in London on September 16, 1932, after being ill for several years. A year after his death, his institute was incorporated into the London School of Hygiene and Tropical Medicine.

Although much work is still to be done, Ross's discovery has allowed for immense progress in combating malaria. Historically, from northern Australia to southern England, malaria was prevalent in most regions of the world. As the University of Oxford's website Our World in Data states, "poet Friedrich Schiller contracted the disease in Mannheim, Oliver Cromwell in Ireland, and Abraham Lincoln in Illinois." Throughout the 20th century, Ross's work meant that malaria was eliminated across many regions of the world, which saved millions of lives. Today, malaria is present almost exclusively in the tropics and, while there is still a long way to go to before the total elimination of the disease, incidences of malaria continue to fall throughout the world. The World Health Organization estimates that malaria deaths dropped by more than 47 percent between 2000 and 2015. Without Ross's work, the progress in combating malaria would be far less sophisticated, and tens, if not hundreds, of millions of people would not be alive today.

22

Karl Landsteiner and Richard Lewisohn

Pioneers of blood transfusions

Austrian physician and immunologist Karl Landsteiner and German surgeon Richard Lewisohn were pioneering European scientists in the field of blood transfusion. Landsteiner discovered the existence of different blood groups and realized that transfusions between people of identical or compatible blood groups did not lead to the destruction of blood cells. Lewisohn developed procedures that allowed blood to be stored outside the body without clotting. Together, the pair laid the foundation for the modern field of blood transfusion medicine and the worldwide system of blood banks that we have today. By making blood transfusions much safer and more practical, their work is credited with saving about one billion lives.

Karl Landsteiner was born on June 14, 1868, in an upper-middle-class suburb of Vienna, Austria. He was raised by his mother. His father, the editor in chief of a popular Austrian newspaper, died when Karl was six years old. Landsteiner excelled as a student and enrolled in medical school at the University of Vienna in 1885, at the age of 17, to study anatomy, blood chemistry, zoology, and organic chemistry.

In 1888, Landsteiner took a break from his studies to complete a year of mandatory military service. Nevertheless, he graduated with his medical degree in 1891 at the age of 23. Notably, while still a student, Landsteiner published a paper on the impact of diet on blood composition. After graduation, he continued his studies in microbiology and spent five years visiting various European universities.

In 1896, Landsteiner returned to Austria and became a research assistant at Vienna's Hygiene Institute. Two years later, he was hired as an assistant at the University of Austria's Department of Pathological Anatomy. During his 10 years on the job, Landsteiner performed more than 3,600 autopsies, and it was in Vienna where he made his world-changing discovery.

Since the mid-1870s, scientists had known that if humans are given blood transfusions from other animals, the foreign animal's blood usually causes agglutination, meaning red blood cells clump and potentially release lethal toxins into the patient's bloodstream. In 1900, in a footnote in a paper on a different topic, Landsteiner noted that agglutination of blood might also happen when the blood of one human is transfused to another.

Landsteiner believed this agglutination caused the complications that often followed blood transfusion attempts.

Through repeated trials of combining the blood serum of one person with that of another, Landsteiner found that in some cases, large clumps of red blood cells would form. Yet in other cases, there were only small, nonproblematic clumps. This finding led him to conclude that not all human blood is the same.

In 1901, Landsteiner expanded on his observations in a paper titled "Agglutination of Normal Human Blood." In the article, he declared that, contrary to popular belief, his research had shown "differences of a rather striking kind between blood serum and corpuscles of different apparently entirely healthy human beings." Landsteiner explains that he identified three antigens in samples of human blood and that when these samples are mixed, agglutination occurs.

Once Landsteiner showed that agglutination does not occur when blood samples of the same type are mixed, he classified the blood types as A, B, and C (later designated O). In 1902, a fourth blood type, AB, was discovered by Landsteiner's colleagues, further legitimizing his work.

The discovery of blood types was quickly popularized across the scientific community. This finding meant that blood transfusions became much safer. However, one major drawback remained. Blood transfusions still used the so-called direct method, in which the donor and the recipient had to be side by side for the transfusion to be given. This made transfusions impractical, as it was difficult to find a donor who had the correct blood type and could come to the patient. Fortunately, this is where Richard Lewisohn enters the story.

Richard Lewisohn was born in Hamburg, Germany, on June 12, 1875. He attended the local *gymnasium*, a type of secondary school in the German education system that strongly emphasizes academic learning. In 1893, he enrolled in medical school and, as was typical for German medical students at the time, he attended several different universities. Lewisohn received a doctorate from the University of Freiburg, in southwest Germany, in 1899.

After graduating, Lewisohn worked as an assistant for two prominent German pathologists before emigrating to New York in 1906, where he worked as a surgeon at Mount Sinai Hospital. Landsteiner's discovery of different blood types in the early 1910s renewed interest in blood transfusion.

The goal of making the procedure safer and more widely applicable greatly interested Lewisohn. His difficult challenge was to find a way to store blood outside the body without the blood clotting. Any clotting (coagulation) would render the blood useless for transfusions.

Lewisohn built on the work of Belgian physician Albert Hustin, who in 1914 proved that sodium citrate could be added to blood as an anticoagulant. In 1915, after continuous experimentation, Lewisohn discovered that the optimal concentration of sodium citrate equaled 0.2 percent of the total mass of blood without exceeding 5 grams per transfusion. Lewisohn's optimization allowed blood to be stored safely for two days before transfusion.

Lewisohn's method was generally accepted after he published his findings in 1915. Others optimized Lewisohn's method the following year and pushed the time frame for which blood could be stored to 14 days. Today, blood can last upwards of 42 days outside the body if it is kept at a cool temperature and nontoxic preservatives are added. Doctors around the world immediately began using the new procedure, which, coupled with Landsteiner's discovery of blood groups, helped save thousands of lives within just a few years.

After discovering blood groups, Landsteiner became a professor of pathological anatomy at the University of Vienna in 1911. After a two-year stint at a hospital in the Netherlands, he accepted a post at the Rockefeller Institute (now Rockefeller University) in New York City in 1922. At Rockefeller, Landsteiner continued his work on blood groups, creating a more detailed model of which groups can be paired together. He also helped to identify the causative agent of syphilis and made contributions to the scientific community's understanding of polio.

In 1930, Landsteiner received the Nobel Prize in Physiology or Medicine for his discovery of blood groups. Although he formally retired in 1939, he continued his research. On June 24, 1943, in his lab with a pipette in hand, he had a heart attack. He passed away two days later, on June 26, 1943.

Lewisohn remained at Mount Sinai for most of his life. After he retired from active surgery in 1937, he spent the next decade working on cancer research. His endeavors proved successful, and he was the first person to identify the significance of folic acid in cancer.

Lewisohn received several awards and fellowships in recognition of his work. In 1916, he was elected a fellow of the American College of Surgeons and served on the American Board of Surgery. In 1955, he received the American Association of

Blood Banks' Karl Landsteiner Memorial Award. Despite the accolades, Lewisohn modestly described his blood transfusion method as "not in the least complicated." He died on August 11, 1961, in New York City, at the age of 86.

Before Landsteiner's and Lewisohn's breakthroughs, people regularly bled to death from ulcers, accidents, and childbirth. If blood transfusions were given, most patients received an incompatible blood type and died when their bodies rejected the donor's blood. Landsteiner and Lewisohn's work helped make blood transfusions far safer and more practical. Their work led to the establishment of the worldwide system of blood banks, which is credited with saving approximately one billion lives to date.

23

Fritz Haber and Carl Bosch

Chemists who revolutionized agriculture

Fritz Haber and Carl Bosch were 19th-century German Nobel Prize–winning scientists. Together, they created the Haber-Bosch process, which efficiently converts nitrogen from the air into ammonia (a compound of nitrogen and hydrogen). Ammonia is then used as a fertilizer to increase crop yields dramatically. The impact of Haber and Bosch's work on global food production transformed the world forever.

Throughout the 19th century, farmers used guano (the accumulated excrement of seabirds and bats) as an extremely effective fertilizer due to its exceptionally high content of nitrogen, phosphate, and potassium—nutrients essential for plant growth. However, by the beginning of the 20th century, guano deposits around the world had begun to run out, and the price of fertilizer began to increase rapidly. If a solution to the depletion of guano hadn't come soon, global famine would have followed.

Fritz Haber was born on December 9, 1868, in what was then Breslau, Germany, but is today Wroclaw, Poland. Haber's mother passed away just three weeks after he was born, leaving Fritz to be raised jointly by his father and various aunts. Haber's father was a successful merchant, and Fritz's interest in chemistry from an early age was in part due to his father's work selling natural dyes and pigments.

Between 1886 and 1891, Haber studied chemistry at the University of Heidelberg, the University of Berlin, and the Technical School at Charlottenburg. During this time, he also completed a year of mandatory military service. In 1891, Haber received his doctorate.

After graduation, Haber spent three years at various academic and industrial jobs, including a brief stint working for his father's chemical business. Haber's father wanted him to become a salesman and learn more about business. Haber didn't agree, and in 1894 he began working as an assistant at the University of Karlsruhe. He published dozens of papers and books on thermodynamics and electrochemistry, and in 1906 became a professor of physical chemistry and electrochemistry.

While at the University of Karlsruhe, Haber became interested in researching methods of synthesizing ammonia from nitrogen. Nitrogen is very common in the atmosphere—78 percent by volume. Still, the element is difficult to extract from the air and turn into a liquid or solid form, a process known as "fixing nitrogen."

After thousands of experiments over 10 years, in 1908, Haber showed that it was possible to synthesize ammonia from hydrogen and atmospheric nitrogen under high pressure and temperature if there was a suitable catalyst. In March 1909, he demonstrated to his colleagues that he had found a way to fix nitrogen. This breakthrough proved that commercial production was possible.

However, there was a problem with Haber's process: it took place in a small tube about 75 centimeters (29.5 inches) tall and 13 centimeters (5 inches) in diameter. In 1909, large containers that could handle the pressure and temperature required for industrial-scale ammonia production did not exist.

That is where Carl Bosch comes in.

Carl Bosch was born on August 27, 1874, in Cologne, Germany. His father, Carl Friedrich Alexander Bosch, and his uncle, Robert Bosch, pioneered the development of commercially available spark plugs and founded Bosch, a large multinational engineering and technology company. Although the company wasn't a worldwide brand in the late 19th century, like it is today, Carl grew up relatively wealthy.

In 1894, Bosch studied metallurgy and mechanical engineering at a technical institute in Charlottenburg before transferring to the University of Leipzig in 1896 to study chemistry. He graduated with a doctorate in chemistry in 1898.

In April 1899, Bosch began working as a chemist for the large, multinational chemical company Badische Anilin- und Sodafabrik, better known by its initials, BASF. In his first few years there, he focused mainly on developing synthetic indigo dye.

In 1908, Bosch found out about Haber's laboratory breakthrough and decided to take on the challenge of developing containers that could withstand Haber's process on the industrial level. After doing more than 20,000 experiments, constructing an entire plant, and making an exhaustive search for a cheaper and more abundant catalyst, Bosch and his team eventually succeeded in synthesizing ammonia on an industrial scale. The Haber-Bosch process was born.

By 1912, Bosch was using 8-meter-tall containers (about 24 feet tall), and in 1913, BASF opened the first factory dedicated to ammonia production, which kick-started the fertilizer industry we know today.

The development of the Haber-Bosch process meant that, for the first time in human history, it was possible to produce synthetic fertilizers that could be used on enough crops to sustain the Earth's growing population. It is nearly impossible to say how many lives this breakthrough helped save. Still, the explosion of the world's population from 1.6 billion in 1900 to about 8 billion today "would not have been possible without the synthesis of ammonia," according to Czech-Canadian scientist Vaclav Smil.

After their revolutionary contribution to human progress, Haber and Bosch did work during World War I that was terribly harmful to humanity. (As noted in the introduction to this volume, inclusion of an individual in this collection does not constitute an endorsement of *everything* that person did.) Haber was a German patriot, and in 1914, he wholeheartedly committed himself and his lab to the war effort. Haber began developing chemical weapons. In 1915, Haber discovered chlorine gas, which was first used in the Second Battle of Ypres. Chlorine gas was used for the remainder of World War I and has been used in many wars since. By 1916, Haber was acting chief of Germany's Chemical Warfare Service. Bosch, meanwhile, worked with his employer, BASF, to manufacture the nitric acid necessary to make saltpeter explosives.

In retrospect, Haber and Bosch's work can be summed up as being responsible for tens, if not hundreds, of thousands of deaths—but hundreds of millions of lives saved.

Haber continued to be involved in chemical weapon development after the war; up until 1923, he helped Russia and Spain develop their gas warfare arsenal. In the mid-1920s, intending to help Germany meet its war reparations, Haber pursued a six-year project that attempted to extract gold from seawater. The project was a failure, and Haber blamed himself for its lack of success.

When Adolf Hitler came to power in 1933, Haber was ordered to dismiss all Jewish personnel from the Kaiser Wilhelm Institute, which he was running at the time. Haber was stunned by these developments and attempted to delay firing his Jewish staff until they could find alternative employment. In October 1933, Haber resigned as director of the institute and as a professor, stating that although he had converted from Judaism and would be legally entitled to stay in his position, he no longer wished to do so. Haber fled to England in late 1933. In early 1934, he accepted a job at a research institute in Mandatory Palestine. On his way there, ill health overcame Haber, and he died in a hotel in Basel, Switzerland, on January 29, 1934.

After the war, in 1919, Bosch was appointed managing director of BASF, and in 1925, he helped found and subsequently became the first head of IG Farben, then the world's largest chemical company. Despite his managerial responsibilities, Bosch continued his experiments with high-pressure catalytic processes, and he successfully produced artificial benzene—a chemical used to make plastics, pesticides, resins, dyes, and more—by hydrogenating coal. Bosch became chairman of IG Farben's board of directors in 1935. In 1937, Bosch was appointed president of the Kaiser Wilhelm Institute, Germany's highest scientific position.

Bosch's attitude toward the Nazis was mixed. Although he was a member of the German Democratic Party and vehemently opposed Hitler's treatment of Jewish scientists, he profited from the National Socialists' economic policies. As head of IG Farben, he entered into contracts with the Nazi government to manufacture synthetic benzene. Bosch's calls for the Nazis to halt the persecution of Jewish scientists in 1930s Germany were met with hostility. The government declared him persona non grata, and he fell into depression and alcoholism. By the late 1930s, he was gradually relieved of his positions, and he died in Heidelberg, Germany, on April 26, 1940. After his death, IG Farben, the company Bosch founded and directed, went on to play an infamous role in Hitler's war machine. The company manufactured Zyklon B, the poison used in gas chambers in the concentration camps; exploited upwards of 35,000 slave laborers, many of whom were imprisoned in Auschwitz; conducted medical experiments on those inmates; and created many other harmful chemical products.

Haber and Bosch were awarded numerous honors for their work in creating the Haber-Bosch process. Despite criticism for his role in World War I, Haber was awarded the Nobel Prize in Chemistry in 1919. Throughout the 1920s and 1930s, Haber became an honorary member of numerous scientific societies. In 1930, Bosch was also awarded the Nobel Prize in Chemistry.

Although the Haber-Bosch Process consumes about 1 percent of the world's energy, it is responsible for feeding more than one-third of the global population. Today, more than 159 million tons of ammonia are produced annually. Although ammonia is also used for cleaning and as a refrigerant, 88 percent of the world's ammonia production goes into the fertilizer industry. Vaclav Smil estimates that if average crop yields had remained at their 1900 level, the crop harvest in the year 2000 would have required nearly four times more cultivated land than was actually needed. That amounts to an area equal to almost half of all land on every continent except Antarctica—rather than just the 15 percent required today.

Despite the controversies and legitimate moral concerns surrounding their work, without Fritz Haber and Carl Bosch, the world's population would be much smaller and the environment much more affected than it is today. On balance, the two truly changed the world for the better.

24
Willis Haviland Carrier

Inventor of modern air conditioning

Willis Haviland Carrier was an American engineer who created the first modern air-conditioning unit. In addition to providing us respite from the summer heat, Carrier's invention has enabled our species to inhabit previously inhospitable places, increased productivity at factories and offices around the world, and saved millions of people from suffering heat-related deaths.

Willis Carrier was born on his family's farm in Angola, New York, on November 26, 1876. He attended Central High School in Buffalo and in 1897 won a four-year state scholarship to attend Cornell University. In 1901, Carrier graduated from Cornell with a bachelor of science in engineering, specializing in electrical engineering. That same year, he began working as a research engineer at the Buffalo Forge Company, which designed and manufactured steam engines and pumps.

Carrier spent the first few months of his new job working on a heating system to dry lumber and coffee. In 1902, Sackett & Wilhelms Lithographing & Publishing Company asked Buffalo Forge to devise a system to control humidity at its printing plant. The high temperatures and humidity levels meant that the printing paper would often soak up moisture from the air, which caused the paper to expand. That was a problem, because the colors used in the printing process became misaligned when the paper changed its size, ruining the product.

Carrier decided to tackle this problem. By doing so, he ended up creating the world's first air-conditioning unit in 1902. Carrier's invention controlled the temperature, humidity, and air circulation while cleaning the air at the Sackett & Wilhelms printing factory. It worked by drawing in air through a filter, passing the air over coils filled with coolant, and then venting the newly cooled and dehumidified air back out. That year, the New York Stock Exchange became the first building to be air-conditioned for comfort. On January 2, 1906, Carrier was issued with a patent for an "apparatus for treating air."

In 1915, after the Buffalo Forge Company decided to focus solely on the manufacture rather than the design of new products, Carrier and six other engineers pooled their life savings of $32,600 (about $986,000 in 2023 dollars) to create the Carrier Engineering Corporation. With his new company, Carrier began to expand the use of air-conditioning units by supplying hotels, department stores, movie theaters, and private homes.

His units were installed in the White House, the U.S. Congress, and the old Madison Square Garden (a predecessor to today's modern arena of the same name).

After experiencing financial problems caused by the Great Depression, Carrier's corporation merged with Brunswick-Kroeschell Company and York Heating & Ventilating Corporation (a division of York Ice Machinery, which was later renamed York International Corporation) to form the Carrier Corporation, with Carrier as board chairman. He spent the rest of his life improving the design and functionality of his air-conditioning units. He died on October 7, 1950, in New York City.

Carrier passed away before he could witness the immense surge in popularity of air conditioning during the postwar economic boom of the 1950s, which saw his invention quickly spread across the United States and to other parts of the world. Thanks to Carrier, for the first time in their history, humans were able to consistently and accurately control the atmosphere inside buildings.

As University of Rochester economist Walter Oi wrote in 1996, in machine shops, labor productivity is at its peak at 65 degrees Fahrenheit with humidity between 65 and 75 percent. Productivity is 15 percent lower at 75 degrees Fahrenheit and 28 percent lower at 86 degrees Fahrenheit. Moreover, work accident rates are 30 percent higher at 77 degrees Fahrenheit than at 67 degrees Fahrenheit. The introduction of air conditioning, he argued, made value added per employee in manufacturing in the American South increase from 88.9 percent of the national average in 1954 to 96.3 percent of the national average in 1987.

Even more striking, Oi noted, was the impact of air conditioning on U.S. mortality rates, which used to be higher in summer and winter than in spring and fall and much higher in the South than in the North. In 1951, the infant mortality rate in the South was 45 percent higher than that in New England; by 1990, it was only 13 percent higher.

In 1942, Carrier was awarded an honorary doctor of letters degree by Alfred University in upstate New York. That same year, he was awarded the Frank P. Brown Medal, a prize for excellence in engineering and science. In 1985, Carrier was posthumously inducted into the National Inventors Hall of Fame. Since its creation, air conditioning has saved and improved millions of lives.

25

Alexander Fleming

Discoverer of penicillin

Sir Alexander Fleming was a 19th-century Scottish scientist who discovered penicillin, a group of beta-lactam antibiotics used to treat a range of infections. His finding paved the way for the development of antibiotic drugs, which have been credited with saving more than 80 million lives to date.

Alexander Fleming was born on August 6, 1881, in Ayrshire, Scotland, and was the seventh of eight children. His father was a farmer. At the age of 13, Fleming moved to London to live with his older brother and attended the Regent Street Polytechnic. He spent four years working as a shipping clerk. After inheriting some money from an uncle at the age of 21 and obtaining a scholarship, he enrolled at St. Mary's Hospital Medical School, London.

Fleming graduated with distinction with a bachelor of medicine and surgery in 1906. At first, he planned to become a surgeon, but after taking a temporary position in the research department at St. Mary's, he became convinced that his future lay in bacteriology. At St. Mary's, Fleming assisted Sir Almroth Wright, a pioneer in vaccine therapy and immunology. In 1908, he graduated with a bachelor of science in bacteriology, was honored as the top medical student at the university, and became a lecturer at St. Mary's.

Fleming had been a member of the London Scottish Regiment of the Volunteer Force, a part-time army reservist unit, since 1900. This meant he was among the first to be called up in 1914 when World War I began. Fleming served in the Army Medical Corps and, for most of the war, studied wound infections in a laboratory inside a military hospital in Boulogne-sur-Mer, France. He was promoted to captain in 1917.

In 1918, he returned to St. Mary's to work as a lecturer. In 1921, he discovered that the enzyme lysozyme, which is present in saliva, tears, and mucus, had mild antiseptic effects. Serendipitously, Fleming made this discovery when he had a cold and a drop of nasal mucus fell into a culture plate of bacteria. When he realized his mucus might have influenced the bacterial growth, he mixed mucus into a culture and a short time later saw that some of the bacteria had been dissolved. This discovery was an important contribution in understanding how the body fights infections. Fleming later remarked that he considered this to be his best work as a scientist.

On September 3, 1928, shortly after he became a professor of bacteriology, Fleming returned to his lab, having spent

August on a holiday with his family. Fleming was notorious for his messy lab. Upon returning from his long vacation, he discovered that he had left out petri dishes of *Staphylococcus*, a common bacterium found in 25 percent of healthy people.

Upon investigation, he noticed that mold had accidentally developed in the plate of staphylococci culture. Interestingly, the mold had inhibited the growth of the bacteria and created a bacteria-free circle around itself. At first Fleming called the substance "mold juice" and decided to investigate it further, as he thought it might be a more potent enzyme than lysozyme.

It turned out that the mold Fleming had discovered was not an enzyme, but the first antibiotic, which he identified as being from the genus *Penicillium*. He conducted more experiments and found that even when diluted 800 times, the mold culture prevented the growth of staphylococci. Further study revealed that penicillin destroys bacteria by attacking their cell walls and interfering with the cells' ability to reproduce. Fleming also found that penicillin could fight all gram-positive bacteria (bacteria with a more penetrable cell wall), including those that cause diphtheria, meningitis, scarlet fever, and pneumonia.

Fleming published his discovery in 1929 in the *British Journal of Experimental Pathology*. However, his findings got little attention at the time. He continued with his experiments but found that cultivation of penicillin was difficult. After growing the mold, isolating the antibiotic agent proved arduous. Lacking the funds and manpower needed for more in-depth research, Fleming was unable to stabilize and purify penicillin and ultimately abandoned his pursuit after a series of inconclusive experiments.

During World War II, American pharmacologist Howard Florey and German biochemist Ernst Boris Chain from the University of Oxford obtained a preserved strain of Fleming's penicillin. They began large-scale research, hoping to be able to mass-produce the antibiotic. Fortunately, mass production began in 1942, and by D-Day in 1944, enough penicillin had been produced to treat wounded Allied troops.

In 1944, Fleming was knighted by King George VI and became Sir Alexander Fleming. That next year, Fleming, Florey, and Chain jointly won the Nobel Prize in Physiology or Medicine for their contribution to developing the antibiotic.

Looking back on the day of his discovery, Fleming said, "One sometimes finds what one is not looking for. When I woke up . . . I certainly didn't plan to revolutionize all medicine by discovering the world's first antibiotic or bacteria killer. But I suppose that's exactly what I did."

Later in life, Fleming was decorated with numerous awards. He was an honorary member of nearly all medical societies across the world, became a Freeman—a mark of distinction given by municipal councils to individuals they wish to honor—of many boroughs and cities, and received honorary doctorates from almost 30 universities across Europe and America. In 1955, he died at age 73 at his home in London.

Fleming's discovery of penicillin laid the foundation for the development of the antibiotic wonder drugs that have been credited with saving over 80 million lives. Penicillin revolutionized the medical field, and most people reading this have likely benefited from Fleming's discovery at some point in their lives.

26

Lucy Wills

Pioneering maternal health researcher

Lucy Wills was an English hematologist who discovered that folic acid can be used to prevent life-threatening types of anemia—a condition in which a person lacks enough healthy red blood cells to carry adequate oxygen to the body's tissues—in pregnant women. Her research into women's health during pregnancy has saved countless lives and improved prenatal care. Today, folic acid is recommended for all pregnant women to help protect the mother from illness and assist in the healthy development of the baby.

Lucy Wills was born on May 10, 1888, in Sutton Coldfield, a town on the outskirts of Birmingham, England. She received a strong scientific education from an early age. Her father was a science graduate, and her mother was the daughter of a well-known doctor. In 1903, Wills attended Cheltenham School, one of the first British boarding schools to train female students in science and mathematics. Four years later, in 1907, she began studying natural sciences and botany at Newnham College, an all-women's college at the University of Cambridge. Wills finished her university examinations in 1911. However, she was ineligible to receive a degree from Cambridge, which did not grant women degrees until 1947.

In 1915, Wills enrolled at the London School of Medicine for Women, the first school in Britain to train female doctors. By 1920, she had become a legally qualified medical practitioner, achieving the requirements for the degrees of bachelor of medicine and bachelor of surgery from the Royal College of Physicians. After graduation, Wills taught and conducted research at the Department of Pregnant Pathology at the Royal Free Teaching Hospital in London.

In 1928, Wills was recruited to work in Mumbai, India. Her task was to investigate why millions of pregnant women in the developing world suffered from a severe and often deadly form of anemia. She found that the red blood cells of the anemic pregnant women were extremely swollen and therefore were not carrying enough hemoglobin, a red protein responsible for transporting oxygen in the blood of vertebrates.

At first, Wills hypothesized that the anemia may have been caused by a bacterium or virus, but after studying the women's living conditions and workspaces, no harmful pathogens were found. She soon realized that wealthier Indian women, who often had a far more nutritious diet, were significantly less likely to become anemic during pregnancy. This led Wills to wonder whether a nutritional deficiency could be the cause of the anemia.

Wills decided to feed rhesus monkeys in a laboratory a diet based on that eaten by the anemic women. Many of the monkeys studied came down with anemia. Wills found that liver extract, which had previously been found to be effective in treating a different form of anemia (Addisonian), was also effective in treating the type of anemia that Wills was researching.

However, Wills also discovered that a much higher dose of the liver extract was needed to combat the anemia afflicting the pregnant monkeys. Since liver extract was expensive at the time, she knew it could not become an effective treatment for low-income anemic women.

A breakthrough came when Wills discovered that anemia could be treated with the addition of yeast extract in the monkeys' diet. Wills found that Marmite, a cheap, popular British breakfast spread made from concentrated brewer's yeast, was extremely successful in treating the anemic monkeys. Wills tested the effectiveness of Marmite and liver supplements on several pregnant anemic women and found that both supplements worked. As liver was far more expensive than the commercially available Marmite, she treated several anemic women with Marmite alone.

In 1931, writing in the *Asia-Pacific Journal*, Wills noted that the improvement in pregnant women's health "was amazing . . . [as] they experienced a quick return of appetite . . . and an increase in the red cell count by the fourth day." Wills published her results even though she acknowledged that she didn't know which compound in the Marmite and the liver extract was responsible for curing her patients.

Scientists from around the world dubbed the unknown compound the Wills factor and encouraged women everywhere to eat Marmite or liver extract during pregnancy. In 1941, the mysterious Wills factor was isolated; today we know it as folic acid.

Wills returned to London to work at the Royal Free Hospital as a full-time pathologist until her retirement in 1948. She spent her later years traveling extensively. Wills continued her research on nutrition and anemia in the developing world. After her death in April 1964, the *British Medical Journal* said that the discovery of the "Wills factor in yeast extract . . . was one of the simple but great observations which are landmarks in the history and treatment of the nutritional anemias."

Today, women everywhere are encouraged to consume folic acid to help ensure a healthy pregnancy. Since its discovery, folic acid has been found not only to prevent anemia in pregnant women, but also to greatly reduce the risk of severe congenital anomalies known as neural tube defects that typically lead to permanent disability, damage to the skull or brain, and often early death.

By discovering a cure for anemia during pregnancy, Wills has prevented the suffering and potential deaths of millions of women and their babies around the world.

27

Pearl Kendrick and Grace Eldering

Creators of the whooping cough vaccine

Pearl Kendrick and Grace Eldering were American scientists who created the first effective vaccine for whooping cough. Thanks to their work, this disease has become preventable, and Eldering and Kendrick's vaccine has been credited with saving more than 15 million lives so far.

Whooping cough is an upper respiratory infection that typically afflicts infants. Although early symptoms are often mild, over time the coughing bouts cause those infected to lose their breath, turn red, and vomit. At the end of a coughing bout, the child will often desperately suck in air, which results in a whooping noise. The disease, which is known as pertussis to scientists, after the bacterium (*Bordetella pertussis*) that causes it, can result in life-threatening complications such as pneumonia, bacterial infections, and dehydration.

At its height during the 1930s, whooping cough killed more American infants than polio, measles, tuberculosis, and all other childhood diseases combined. Fortunately, Kendrick and Eldering helped change this.

Pearl Kendrick was born on August 24, 1890, in Wheaton, Illinois. When she was just three years old, she developed a case of whooping cough. Kendrick was lucky enough to survive the illness and went on to have a happy childhood. She received a bachelor of science in liberal arts from Syracuse University in 1914.

Kendrick began her career as a high school science teacher, but after a short time she started studying bacteriology at Columbia University, focusing on whooping cough. In 1917, she was recruited to work at the Michigan Department of Health, where she would meet Grace Eldering.

Eldering was born in 1900 in Rancher, Montana. Like Kendrick, she contracted and survived whooping cough at a young age. Eldering studied biology and English at the University of Montana and graduated in 1927.

In 1928, Eldering moved to Michigan and began volunteering at the Michigan Bureau of Laboratories. After six months of volunteering, she was placed on the payroll. In 1932, Eldering transferred to the lab Kendrick ran in Grand Rapids.

Kendrick and Eldering instantly bonded and began working together on a whooping cough vaccine. However, as their efforts coincided with the Great Depression, funding for their vaccine was virtually nonexistent. As a result, they developed their vaccine primarily during their off hours. This arrangement worked for a few years, but by 1936 they were in desperate need of additional funds to continue trials for their test vaccine.

To raise funds, Kendrick invited First Lady Eleanor Roosevelt to their laboratory. To everyone's surprise, she accepted the invitation and one day spent more than 13 hours with Kendrick. Soon after her visit, Mrs. Roosevelt helped to find the funds that allowed Kendrick and Eldering to continue the large scale trial they had begun in 1934. The trial eventually came to involve 5,800 children.

The results were groundbreaking. The children who received the vaccine immediately demonstrated a strong immunity against the disease. In 1942, to reduce the discomfort for children receiving vaccines, Eldering and Kendrick combined three vaccines into a single shot. The use of the diphtheria, pertussis, and tetanus, or DPT, vaccine became routine throughout the United States in 1943. Its use then spread quickly around the world.

Both scientists earned PhDs from Johns Hopkins University, Kendrick in 1934 and Eldering in 1942.

Later in life, Kendrick left the Michigan Department of Public Health to teach at the University of Michigan. She died in 1980. After Kendrick left, Eldering succeeded her as the head of the department. Eldering retired in 1969 and died in 1988. In 1983, both women were inducted into the Michigan Women's Hall of Fame.

Tragically, each year 160,000 children in the developing world die after contracting whooping cough. Although this figure continues to fall, there is much to be done before whooping cough is completely eradicated.

However, thanks to Eldering and Kendrick's work, more than 15 million lives have already been saved, and it is likely that their vaccine will continue to save millions more.

28

Abel Wolman and Linn Enslow

Pioneers of chlorination

Abel Wolman and Linn Enslow were 20th-century American scientists who discovered how to safely use chlorine to purify drinking water. They perfected their formula in 1923. Thanks to their work, more than 175 million lives worldwide have been saved so far.

The use of chlorine to purify water did not start with Wolman and Enslow. During a cholera epidemic in 1854, chlorine had been used to purify London's drinking water. The first American patent for a water chlorination system was granted in 1888. Although it was accepted that chlorine could kill bacteria, little was understood about the cleansing process and, because chlorine can be poisonous to humans, using the chemical for water purification remained dangerous.

In the early 20th century, cities across America were expanding at a rapid pace, and advances such as indoor running water were becoming more widespread. With no safe or effective measures to clean their drinking water, city water suppliers often became unwitting disseminators of an array of diseases, including cholera, dysentery, and typhoid. This is where Wolman and Enslow enter the story.

Abel Wolman was born in June 1892, in Baltimore, Maryland, and was one of six children in a Polish-Jewish immigrant family. Although he had initially wanted to go into medicine, his parents encouraged him to study engineering. In 1915, Wolman became the fourth person to receive a bachelor of science from Johns Hopkins University's newly established engineering school.

Linn Enslow was born in February 1891, in Richmond, Virginia. He studied chemistry at Johns Hopkins University, where he met Wolman. After graduation, both Enslow and Wolman began working at the Maryland Department of Public Health. In 1918, they teamed up to study chlorine's effect on water purification.

In developing their method of water purification, Enslow and Wolman analyzed chlorine's effect on the acidity, bacterial content, and taste of drinking water. By 1923, they had created a standard formula detailing the amount of chlorine needed to safely purify water supplies. Their rigorous scientific research laid the foundation for global water purification.

After their breakthrough, Wolman took a more active role than Enslow in encouraging states and countries to adopt the formula. Eventually, Wolman was able to apply the new purification method to Maryland's drinking water supply. By 1930, typhoid cases in the state had declined by 92 percent. By 1941, 85 percent of all U.S. water systems used the Enslow-Wolman formula. The rest of the world followed America's lead.

Wolman's career flourished. He became chair of the state planning commission in his early thirties, acted as a consultant to the U.S. Public Health Service, was chief engineer at the Maryland Department of Public Health, and established the Department of Sanitary Engineering at Johns Hopkins University in 1937.

Throughout his life, Wolman sat on numerous boards and advised governments throughout the world on water purification systems. He eventually retired, in 1962. Wolman died in 1989 in his native Baltimore at the age of 96.

Enslow went on to become the editor of the journal *Water and Sewage Works.* He worked in that role until his sudden death from a heart attack in 1957, at his farm in Virginia.

Thanks to the work of Enslow and Wolman, billions of people now have access to drinking water that is free from the etiologic agents of an array of potentially deadly diseases. It is estimated by the World Economic Forum that the adoption of their formula in water systems worldwide has saved almost 200 million lives.

29

Frederick Banting and Charles Best

Diabetes treatment trailblazers

Frederick Banting and Charles Best were North American scientists who created the first effective treatment for diabetes by successfully extracting the hormone insulin from the pancreas. Thanks to Banting and Best's work, millions of diabetics can now live long, healthy lives rather than face early, painful deaths.

Diabetes causes a person's blood sugar levels to become too high. Its symptoms include excessive thirst, nausea, fatigue, sugary urination, and weight loss. If left untreated, it can lead to complications that include strokes, kidney failure, heart attacks, and nerve damage. Diabetes has plagued humanity for thousands of years, but even a century ago there were still no effective treatments. Enter Frederick Banting and Charles Best.

Frederick Banting was born on November 14, 1891, in his family's farmhouse near Alliston, Ontario. In 1912, he began studying medicine at Victoria College, part of the University of Toronto. He joined the Canadian army in 1915 and graduated a year later. In 1918, during World War I, he was wounded at the Battle of Cambrai and in 1919 was awarded the Military Cross for heroism under fire.

After the war, Banting returned to Canada and studied orthopedic medicine. From 1919 to 1920, he was resident surgeon at the Hospital for Sick Children in Toronto. In 1921, he began lecturing in pharmacology at the University of Toronto. During this time, Banting became interested in the study of diabetes.

Before the 1920s, it was known that diabetes resulted from a lack of a hormone called insulin, which is made in the pancreas. It was thought that insulin controlled the body's metabolism of sugar. Lack of insulin, people believed, led to an increase of sugar in the blood.

Unfortunately, previous attempts to extract insulin from the pancreas had failed because trypsin, a digestive enzyme produced by the pancreas, would break down the pure insulin before it could be extracted. Banting had to find a way of extracting the insulin from the pancreas before it could be destroyed by the organ's own digestive enzyme.

Banting read about a 1920 experiment by Moses Barron, a Russian-American scientist who closed the pancreatic duct and found that the cells that secreted trypsin deteriorated but the cells in the pancreas that are responsible for the production and release of insulin remained intact. This led Banting to theorize that if the pancreatic duct were closed and the trypsin-secreting cells died, insulin could be extracted from the pancreas and given to diabetics.

In the spring of 1921, Banting visited John James Rickard Macleod, a professor of physiology at the University of Toronto, to talk over his theory. After a lengthy discussion, Macleod agreed to give Banting laboratory space and 10 dogs on which to experiment and appointed Charles Best as Banting's assistant.

Charles Best was born in West Pembroke, Maine, on February 27, 1899. In 1915, he began studying physiology and biochemistry at the University of Toronto. He enlisted in the Canadian army in 1918 and, after the war, completed his degree in 1921. The same year, he began studying at the University of Toronto's medical school.

Banting and Best started working together and were quickly successful in isolating insulin from the test dogs' pancreases. After injecting the insulin into dogs whose pancreases had been removed, they found that the dogs who were suffering from their artificially induced diabetes quickly recovered.

Animal insulin is safe for human use, and Banting and Best began taking insulin from the larger pancreases of cows. However, when they encountered problems in refining the insulin solution, Macleod hired James Collip, a professor of biochemistry at the University of Alberta, to work on the purification of insulin.

In January 1922, Banting and Best administered purified insulin into their first patient, Leonard Thompson, a 14-year-old diabetic who was close to dying. Best and Banting's insulin proved to be a success, as Thompson regained his health. The use of insulin to treat diabetes quickly spread around the world.

Banting received his MD in 1922. In 1923, Banting and Macleod were jointly awarded the Nobel Prize in Physiology or Medicine. Banting was unhappy that Macleod who, in Banting's view, had contributed nothing more than resources, was awarded the prize. As a result, Banting split his prize money with Best. Macleod similarly split his prize money with Collip.

In 1923, Banting was elected chair of the new Banting and Best Department of Medical Research, which was endowed by the Legislature of the Province of Ontario. His research focused on silicosis and cancer. In 1925, Best was awarded an MD and, in 1929, he succeeded Macleod as professor of physiology at the University of Toronto.

In 1938, Banting started working for the Royal Canadian Air Force, researching the physiological problems encountered by pilots flying high-altitude aircraft. On February 21, 1941, he died of wounds following the crash of an aircraft in which he was a passenger. Following Banting's death, Best took over as director of the Banting and Best Department of Medical Research. Best spent most of his career investigating carbohydrate metabolism. He retired in 1965 and died on March 31, 1978, in Toronto.

Banting and Best received numerous awards and honorary degrees throughout their lives. Both were members of many medical academies. In 1994, they were inducted into the Canadian Medical Hall of Fame. In 2004, both men were inducted into the National Inventors Hall of Fame, which honors holders of U.S. patents for significant technological advances.

Thanks to the work of Frederick Banting and Charles Best, diabetes went from an untreatable condition that has killed untold millions of people over thousands of years to a disease that can be treated, enabling diabetics to lead normal, healthy lives.

30

Frederick McKinley Jones

Inventor of mobile refrigeration

Frederick McKinley Jones was an American inventor, engineer, and entrepreneur. With more than 60 registered patents in various fields, he was one of the most prolific African American inventors of the 20th century. Jones is best known for developing mobile refrigeration systems for trucks, trains, and ships. His work meant that fresh produce and other perishable goods could be delivered anywhere on a large scale without spoiling, regardless of the season. Starting in World War II, Jones's refrigeration units were also used to transport blood, organs, and vaccines worldwide. Newer versions of his refrigeration units are still in use and have been used extensively to transport COVID-19 vaccines. Mobile refrigeration revolutionized the supermarket and restaurant industries, leading to billions of people being better fed, and it transformed the medical industry, helping to save millions of lives.

Jones was born on May 17, 1893, in Covington, Kentucky. His mother left while he was young, and his father, a railroad worker, struggled to raise him alone. When Jones was seven years old, his father sent him to live with a priest in Cincinnati, Ohio. However, at the age of 11, just after finishing sixth grade, Jones left school and ran away from the priest. Jones ended up taking odd jobs across Cincinnati, and while working as a garage janitor, he discovered a passion for automobile mechanics. Despite his lack of formal education, Jones observed the mechanics and absorbed as much information as possible. Within three years, he became the garage's foreman.

In 1912, after short stints working at a hotel and aboard a steamship, Jones moved to Hallock, Minnesota, and began working as a mechanic on a large farm. During this time, he started building racecars to drive at county fairs and racing exhibitions. Jones's cars were designed and built so well that they overwhelmed the competition, and he became one of the most well-known racers in the Great Lakes region. In 1913, Jones secured an engineering license.

During World War I, Jones joined the U.S. Army as an electrician, and while serving in France, he performed the necessary wiring to ensure that his camp was equipped with telegraphs, electricity, and telephones. He was discharged with the rank of sergeant in 1919 and returned to the farm in Hallock.

Soon after his return, Jones built a transmitter for the town's first radio station. He also regularly helped doctors by driving them to house calls during the harsh Minnesota winter.

When the snow became so deep that it became impossible to drive a car, Jones attached skis to an airplane fuselage, along with a propeller and motor. Soon doctors were zipping around Kittson County at high speeds in Jones's "snowmachine." Although he never patented it, in effect, Jones had built an early version of the snowmobile. Similarly, when he heard a local doctor lamenting that patients had to travel to the clinic for x-rays, Jones developed a portable x-ray machine that doctors could take to patients' homes.

In the mid-1920s, Jones invented the process and devices that enabled silent-movie projectors to play recorded sounds. For the first time, "talking pictures" were possible. This invention attracted the attention of local businessman Joseph A. Numero, who owned a company that developed audio equipment. Numero hired Jones in 1927, and for several years Jones concentrated on converting silent-movie projectors into talking ones. Jones also found ways to stabilize and improve the picture quality of projectors. In the 1930s, he invented and patented a machine for movie theaters that automatically dispensed tickets and change to customers.

However, Jones's most important invention resulted from a six-dollar bet made by Numero during a game of golf. One of Numero's friends, who owned a trucking business, complained that he had a contract to transport raw chicken from Saint Paul, Minnesota, to Chicago, Illinois—a 400-mile drive—but because of high temperatures, an entire load of chicken had spoiled. Numero responded that his engineer, Jones, could easily solve that problem and create a refrigerated trailer in just 30 days. Numero's friend was skeptical, and they made a bet for six dollars (about $125 in 2023 dollars).

Within two weeks, Jones had designed a prototype, and within 30 days, he had a working model for the first unit, called Thermo Control Model A. Jones's design attached refrigeration equipment to a truck's undercarriage, and chilled air flowed into the trailer via refrigerant tubing.

Numero immediately recognized the potential of this invention and promptly sold off his audio equipment business, Cinema Supplies Inc.—which owed its success in large part to Jones's inventions and improvements—to RCA. In 1938, he formed a partnership with Jones called the U.S. Thermo Control Company (renamed Thermo King Corporation in 1941), with Jones as vice president. The same year, Jones filed a patent for the Model A refrigeration unit, which he received in 1949.

Jones then modified the design so it could be fitted on trains and ships, and by 1941, he had created the Model C. This newer model mounted the refrigeration unit on the front of the truck and was lighter and more durable than previous designs. The Model C was manufactured for military use during World War II and became critically important for preserving medicine, blood, and food for army hospitals and troops on the front lines. The U.S. military installed Jones's invention on trucks, boats, and planes. Jones also developed cutting-edge refrigerators for military field kitchens and air-conditioning units for field hospitals.

After the war, the Thermo King company grew rapidly. In the 1940s, Jones came up with gasoline-powered refrigerated boxcars, which helped further reduce shipping costs and made fresh produce more widely available and cheaper. By 1949, Thermo King was valued at $3 million (about $38.5 million in 2023 dollars). In the 1970s, the company expanded to Europe.

Today, it continues to sell newer versions of Jones's invention worldwide.

Jones accumulated more than 60 patents in various fields, including for refrigeration units, engines, sound equipment, and x-ray machines. He received dozens of awards and honors, both during his lifetime and posthumously. In 1944, Jones became the first African American member of the American Society of Refrigeration Engineers. In 1977, he was inducted into the Minnesota Inventors Hall of Fame, and in 1991, President George H. W. Bush posthumously awarded him the National Medal of Technology. In 2007, Jones was inducted into the National Inventors Hall of Fame.

Jones continued to work for Thermo King Corporation, and during the 1950s, he became a consultant for several branches of the government, including the Bureau of Standards and the Department of Defense. In 1961, at the age of 67, Jones died of lung cancer in Minneapolis.

By inventing practical mobile refrigeration units, Frederick McKinley Jones helped change consumers' eating habits forever. Before the invention of practical mobile refrigeration units, produce usually had to be transported in vacuum-sealed cans. But now, people can eat fresh produce year-round, which has undoubtedly improved the health of billions of people and transformed the global economy. By improving the transport of blood, vaccines, and organs, Jones's invention also transformed the medical industry and in the process has saved millions of lives.

31

Paul Hermann Müller

Slayer of mosquito-borne illnesses

Paul Hermann Müller was a 20th-century Swiss chemist who discovered the insecticidal qualities of dichloro-diphenyl-trichloroethane (DDT). DDT's effectiveness in killing mosquitoes, lice, fleas, and sand flies that carry malaria, typhus, the plague, and some tropical diseases has saved countless millions of lives.

Müller was born on January 12, 1899, in Solothurn, Switzerland. At the age of 17, he left school, and one year later he began working as an assistant chemist for Lonza, one of the world's largest chemical and biotechnology companies.

In 1918, Müller returned to school. In 1919, he started studying chemistry with minors in botany and physics at the

University of Basel. Upon completing his undergraduate degree in 1922, he stayed at Basel to study toward a PhD in organic chemistry. Müller completed his PhD in 1925. That year, he started working for J. R. Geigy, a company that specialized in dyes, chemicals, and pharmaceuticals. That is where Müller made his great discovery.

In 1935, Switzerland began experiencing major food shortages caused by crop infestations, and the Soviet Union experienced the most extensive and lethal typhus epidemic in history. These two events had a profound impact on Müller. Before the 1940s, insecticides were either expensive natural products or inexpensive but made from arsenic compounds that made them poisonous to humans and other mammals.

Motivated by the need to create a cheap and long-lasting plant protection agent that did not harm plants or warm-blooded animals, Müller decided to switch the focus of his work at J. R. Geigy from research on vegetable dyes and natural tanning agents to plant protection agents.

By 1937, Müller had developed a successful seed disinfectant named graminone that protected seeds from soil-borne pathogens and insects. He then turned his attention to insecticides.

After years of intensive work and 349 failed experiments, Müller found the compound he was looking for in September 1939—the same month that Nazi Germany invaded Poland, starting World War II. DDT had first been synthesized by Viennese pharmacologist Othmar Zeidler in 1874. Unfortunately, Zeidler failed to recognize DDT's value as an insecticide. Decades later, Müller found what Zeidler had missed.

Müller's discovery of DDT's effectiveness as an insecticide came at an important moment in history. It played a crucial role in protecting Allied troops in Asia, where the shirts of British and American troops were often doused with the compound.

J. R. Geigy obtained a Swiss patent on DDT in 1940 and British, American, and Australian patents in the early 1940s. In 1943, following the liberation of the region by Allied forces, DDT was used in Naples to bring a typhus epidemic under control in just three weeks. Between the 1950s and the 1970s, DDT was used to eradicate malaria in many countries, including the United States and most of southern Europe.

The use of DDT declined after 1972, when it was banned because of concerns raised by the U.S. Environmental Protection Agency. As Richard Tren and Roger Bate of the public health advocacy group Africa Fighting Malaria have noted, "While there is evidence that the widespread, virtually unregulated agricultural use of DDT in the 1950s and 1960s did harm the environment, no study in the scientific literature has shown DDT to be the cause of any human health problem."

In 2006, the World Health Organization reversed its stance on DDT and now recommends "the use of [DDT in] indoor residual spraying" as "DDT presents no health risk when used properly."

After his discovery, Müller went on to become J. R. Geigy's deputy director of research for plant protection. In 1948, he received the Nobel Prize in Physiology or Medicine. The fact that Müller was presented with this award, even though he was not a physiologist or a medical researcher, highlights the immense impact that DDT had in the fight against disease.

Later in life, Müller received many awards and honorary doctorates. In 1961, he retired from J. R. Geigy but continued doing research in his home laboratory. In 1965, Müller died at the age of 66 in Basel.

Thanks to Müller's work, billions of people have been able to avoid exposure to deadly diseases that have plagued humanity since the dawn of our species.

32

Enrico Fermi

Nuclear energy pioneer

Enrico Fermi was an Italian American physicist who created the world's first nuclear reactor. Although controversial, nuclear power remains the world's main source of zero-carbon energy, which NASA scientists calculate has saved millions of people from air pollution–related deaths. Today, 26 percent of electricity in the European Union and 20 percent of electricity in the United States is generated by nuclear power.

Enrico Fermi was born on September 29, 1901, in Rome, Italy. His father was a division head in the Ministry of Railways, and his mother worked as an elementary school teacher. Even at a young age, Fermi showed a keen interest in science and could often be found building scientific contraptions, such as gyroscopes and electric motors. He was baptized as a Roman Catholic but remained an agnostic throughout his life.

In 1918, Fermi graduated from high school and won a scholarship to the prestigious Scuola Normale Superiore in Pisa, Italy. He initially chose to major in mathematics but soon switched to physics, focusing on quantum mechanics and atomic physics. The faculty were so impressed with Fermi's intellect that they quickly put him in the doctoral program. His academic adviser, Luigi Puccianti, used to say that Fermi was so bright that there was little he could teach him.

Fermi earned a doctorate in physics in 1922, when he was just 20 years old. In 1923, he was awarded a scholarship by the Italian government, which allowed him to spend several months studying with renowned physicist Max Born at the University of Göttingen, in Germany. Fermi also received a scholarship from the Rockefeller Foundation to study at the University of Leyden. He moved back to Italy in late 1924.

In Italy, Fermi was appointed lecturer of mathematical physics and mechanics at the University of Florence, a post he would hold for two years. In 1927, he was elected professor of theoretical physics at the University of Rome. In March 1929, Fermi was appointed a member of the Royal Academy of Italy by Benito Mussolini.

In the early stages of his career, Fermi worked primarily on electrodynamic problems and theoretical investigations into spectroscopic phenomena, specifically the interaction between matter and electromagnetic radiation. In 1934, he began to study the atom. He demonstrated that nuclear transformation could occur in nearly every element subjected to neutron bombardment. When he split the uranium atom, Fermi found that the experiment led to the slowing down of neutrons, which caused nuclear fission and the production of new elements beyond those listed in the periodic table at the time.

In 1938, Fermi was awarded the Nobel Prize in Physics "for his work with artificial radioactivity produced by neutrons, and for nuclear reactions brought about by slow neutrons." At that time, Italy had just passed anti-Semitic laws that threatened Fermi's Jewish wife, Laura, and put many of his research assistants out of work. When Fermi and Laura traveled to Stockholm for the Nobel Prize award ceremony, they decided not to return to Italy. Instead, they traveled with their children, a boy and a girl, to the United States.

Fermi was offered several positions across the United States and accepted a physics professorship at Columbia University in New York. While at Columbia, he found that when uranium neutrons were emitted into another batch of fissioning uranium, they would split the uranium atoms and set off a chain reaction that released a tremendous amount of energy. Fermi worked relentlessly to pursue the idea of nuclear energy and, after moving to the University of Chicago in 1942, he successfully constructed the first artificial nuclear reactor, named Chicago Pile-1.

Built in a squash court situated underneath the University of Chicago's football field, the Chicago Pile-1 was almost 25 feet in diameter. It contained 380 tons of graphite blocks, almost 6 tons of uranium metal, and 40 tons of uranium oxide—all arrayed in a carefully designed pattern. Construction of the reactor was completed on December 1, 1942. The next day, the reactor reached a state in which its nuclear-fission chain reaction became self-sustaining. The experiment was the first controlled nuclear chain reaction. Chicago Pile-1 quickly became the prototype for many other large nuclear reactors that were being built across the United States.

In 1944, Fermi began working as an associate director at the Manhattan Project, which focused on the development of the atomic bomb, at Los Alamos National Laboratory in New Mexico. The same year, Fermi and his wife and children became American citizens. After World War II ended, Fermi accepted a professorship at the University of Chicago and was appointed to the U.S. General Advisory Committee for the Atomic Energy Commission.

For the remainder of his life, Fermi's work focused on high-energy physics. He also led investigations on the origins of cosmic rays. In 1954, Fermi was diagnosed with incurable stomach cancer. He died on November 28, 1954, at his home in Chicago.

Many awards, institutions, and concepts are named after Fermi, including the Fermilab in Illinois, the Enrico Fermi Award given by the U.S. Department of Energy, and the Fermi Gamma-Ray Space Telescope. Fermi is also one of only 16 scientists who have an element named after them. It is called *fermium* (Fm).

Nuclear fission is one of the most significant discoveries in human history. Nuclear reactors have provided humanity with reliable and relatively safe and clean energy for close to eight decades. Accidents have been rare and, apart from a few incidents such as the Chernobyl disaster in 1986—which was the result of a flawed reactor design, inadequately trained personnel, and general Soviet managerial incompetence—manageable in terms of their negative impact on humans and the environment.

Today, nuclear power remains the only reliable source of energy that emits zero carbon dioxide into the atmosphere and can be scaled to meet the growing needs of civilization. Although nuclear power has already improved hundreds of millions of lives, excessively burdensome regulation in most nations regarding the use of this power source puts the future use of this technology at risk. If regulation becomes more moderate and the building of new stations is permitted, this renewable technology could help improve billions of lives worldwide with minimal environmental impact.

33

George Hitchings and Gertrude Elion

Pioneers of rational drug design

George Hitchings and Gertrude Elion were American scientists who pioneered the development of rational drug design. For most of history, the traditional method of drug invention relied on a trial-and-error approach to determine the effectiveness of different treatments. Contrary to the conventional ways, the new method developed by Hitchings and Elion concentrated on studying the differences between human cells and disease-causing pathogenic cells. From these findings, drugs can be designed to specifically target harmful pathogens.

Hitchings and Elion's rational drug design has changed the way many new drugs are developed. Their method has led to the creation of drugs to fight leukemia, malaria, gout, organ transplant rejection, and rheumatoid arthritis, among many other ailments.

George Hitchings was born on April 18, 1905, in Hoquiam, Washington. When he was 12 years old, his father died from a prolonged illness. Hitchings later said that the impact of his father's death made him want to pursue a career in medicine.

Hitchings graduated from Franklin High School in Seattle in 1923. That year, he enrolled at the University of Washington to study chemistry. After graduating cum laude in 1927, Hitchings stayed at the University of Washington and obtained a master's degree in 1928. He then moved to Harvard University as a teaching fellow and received a PhD in biochemistry in 1933.

Over the next decade, Hitchings held several temporary appointments at various institutions. As he later admitted, his career "really began in 1942," when he joined Wellcome Research Laboratories, now part of GlaxoSmithKline, as head of the biochemistry department. Two years later, Hitchings wanted to hire a research assistant, and that is when he first met Gertrude Elion.

Gertrude Elion was born on January 23, 1918, in New York City. She was a bright student who graduated from high school at just 15 years old. The same year, her grandfather died of cancer. Elion later enrolled at Hunter College on a full academic scholarship. As with Hitchings, the death of a loved one led to Elion's lifelong commitment to a career in medicine.

After graduating from Hunter College with a degree in chemistry in 1937, she ran into what she described as a "brick wall." The Great Depression made jobs very difficult to come by, and women looking for jobs in science faced still greater obstacles. Elion later reflected that in the 1930s, few employers took her seriously. After she interviewed for jobs that she was qualified for, interviewers would often say she would be a distraction in a lab full of men.

After being rejected by numerous employers and graduate programs, Elion accepted an unpaid position as a laboratory assistant to a chemist. In 1939, she enrolled in a master's program in chemistry at New York University. Elion worked as a high school teacher throughout her master's studies and obtained her master of science in 1941.

In the early 1940s, with many men in the military, new doors opened for women in the sciences and other fields. After working as an analytical chemist for a food company, Elion became bored. After a six-month stint in a lab run by Johnson and Johnson, she was hired by Hitchings as his research assistant in 1944.

Hitchings thought there must be a more rational approach to drug development than the existing trial-and-error method. Hitchings and Elion developed a method of drug design that focused on determining the differences in biochemistry and metabolism between human cells and disease-causing microbes. The new method allowed them to design specific chemicals that could kill or inhibit the reproduction of pathogens without damaging any healthy human cells.

Using their rational drug design method, Hitchings and Elion successfully developed drugs that are used to treat a variety of conditions, including leukemia, gout, malaria, meningitis, and viral herpes, just to name a few. Researchers around the world quickly copied their approach. Within a few years, scientists using the rational drug design method had created medicines to fight the viruses that cause cold sores, chickenpox, and shingles. Eventually, they developed azidothymidine (better known by its initials, AZT), the first treatment available for human immunodeficiency virus (HIV) or acquired immune deficiency syndrome (AIDS).

Elion later wrote, "When we began to see the results of our efforts in the form of new drugs which filled real medical needs and benefited patients in very visible ways, our feeling of reward was immeasurable."

Later in her career, Elion taught at Duke University and the University of North Carolina at Chapel Hill. In 1967, when Hitchings became vice president in charge of research at Burroughs-Wellcome, Elion succeeded Hitchings in his old job.

In 1976, Hitchings became a scientist emeritus at Burroughs-Wellcome. He also served as an adjunct professor at Duke University between 1970 and 1985. Elion officially retired in 1983, but like Hitchings, she continued working in the laboratory on a part-time basis as a scientist emeritus.

Throughout their lives, Hitchings and Elion received dozens of awards and honors. Most notably, they received the Nobel Prize in Physiology or Medicine in 1988, which they shared with Sir James Black, a Scottish pharmacologist who invented propranolol, a commercially successful beta blocker, and cimetidine, a drug that targets ulcers and suppresses the formation of gastric acid. In 1974, Hitchings became a member of the Medicinal Chemistry Hall of Fame and a Foreign Member of the Royal Society.

In 1991, Elion received the National Medal of Science from President George H. W. Bush. That year, she became the first woman inducted into the National Inventors Hall of Fame for her important work developing a drug called 6-mercaptopurine, which offered new hope in the fight against leukemia. Like Hitchings, Elion also became a Foreign Member of the Royal Society in 1995. Although Elion never received a formal doctorate, she was awarded an honorary PhD from Polytechnic University of New York (today part of New York University) in 1989 and an honorary doctor of science from Harvard University in 1998.

Hitchings died at the age of 92 in 1998, and Elion died the following year at 81. Both passed away at their homes in Chapel Hill, North Carolina.

The work of Elion and Hitchings fundamentally transformed the traditional trial-and-error method of drug discovery. Their rational drug design method has been used to create dozens of different treatments for an array of life-threatening illnesses. Rational drug design has already saved or prolonged untold millions of lives and promises to save millions more as more drug treatments are developed.

34

Virginia Apgar

Inventor of the life-saving Apgar score

Virginia Apgar was an American anesthesiologist and medical researcher who created a test to quickly assess the health of newborn babies and determine whether infants need immediate neonatal medical care. The test, named the Apgar score, continues to be used as a standard practice globally and is credited with saving millions of babies' lives since 1952.

Virginia Apgar was born in Westfield, New Jersey, on June 7, 1909. She had two older brothers, one of whom died at a young age from tuberculosis, while the other lived with a chronic illness. Inspired by both of her brothers' medical problems, she opted for a medical career. In 1929, Apgar earned a degree in zoology with minors in physiology and chemistry from Mount Holyoke College. The same year, she began her medical training at Columbia University's College of Physicians and Surgeons (P&S).

Apgar obtained her MD in 1933 and began a two-year surgical internship at P&S's Presbyterian Hospital. Despite her good performance, P&S's chairman, worried about the economic prospects of new women surgeons during the Great Depression, advised Apgar to make a career in anesthesiology, a new field of study that was beginning to take shape as a medical, rather than strictly nursing, specialty.

Apgar accepted the advice, and after her internship ended in 1936, she began a year-long anesthetist nurse training course at Presbyterian Hospital. After completing the course, she performed residencies in anesthesiology at the University of Wisconsin and Bellevue Hospital in New York City. In 1938, Apgar returned to Presbyterian Hospital and became the director of the newly established division of anesthesia. She was the first woman to hold the position of a director at Presbyterian Hospital.

In 1949, Apgar became the first woman to hold a full professorship at P&S in anesthesiology. That freed her from many administrative duties, enabling her to devote more of her time to research.

Apgar noticed that the mortality of infants (defined as babies up to a year old) in the United States rapidly declined between the 1930s and the 1950s. However, the death rate for babies in the first 24 hours after birth stayed the same. Perplexed by this discrepancy, she began documenting the differences between healthy newborns and newborns requiring medical attention.

In 1952, Apgar created a test called the Apgar score that medical professionals could use to assess the health of newborn infants. The Apgar scoring system gives each newborn a cumulative score of 1 to 10, based on the sum of scores of 0, 1, or 2 in five categories. Zero denotes the worst possible condition, and 2 represents the ideal condition for each category: (a) activity (muscle tone), (b) pulse, (c) grimace (reflex irritability), (d) appearance (skin color), and (e) respiration. To make her assessment easy to remember, the first letter of each category spells "APGAR."

The test is usually performed on newborn babies one minute and five minutes after birth. A cumulative score of 3 or below is typically considered critically low and a cause for immediate medical action. Apgar's test soon became common practice all over the world. It remains a standard procedure to assess the health of newborns.

In 1959, Apgar graduated with a master's in public health from Johns Hopkins University and began working for the March of Dimes Foundation—an American nonprofit organization that works to improve the health of mothers and babies—directing its research program with a focus on the treatment and prevention of birth defects.

While working at the March of Dimes, Apgar became an outspoken advocate for universal vaccinations to prevent mother-to-child transmission of rubella. Later in life, she became a lecturer and then a clinical professor of pediatrics at Cornell University. She died on August 7, 1974.

During her career, Apgar received numerous honorary doctorates, was awarded the Distinguished Service Award from the American Society of Anesthesiologists in 1966, and was named Woman of the Year in Science by *Ladies' Home Journal* in 1973. In 1995, she was inducted into the U.S. National Women's Hall of Fame.

The use of the Apgar score is credited with lowering the infant mortality rate by considerably increasing the likelihood of babies' survival during the first 24 hours after birth. The invention and use of Apgar's test have saved millions of lives and continue to save thousands more every day.

35

Willem Kolff

Inventor of kidney dialysis

Willem Kolff was a Dutch physician who invented the kidney dialysis machine. He also played a role in developing the world's first artificial heart and later, the first artificial eye. The World Economic Forum has estimated that since its invention, Kolff's dialysis machine, or what he liked to call the artificial kidney, has saved more than nine million lives.

Willem Kolff was born on February 14, 1911, in Leiden, the Netherlands, into an old Dutch patrician family. He suffered from dyslexia, but the condition was not recognized at the time, and he was often punished in school for the difficulties he had in reading and spelling. Initially, Kolff wanted to become the director of a zoo, but after his father pointed out that that career path had very limited job opportunities, as there were just three zoos in the Netherlands at the time, he decided to follow in his father's footsteps and pursue a medical career.

Kolff began studying medicine at the University of Leiden in 1936 and was awarded an MD in 1938. Later that year, he began studying for a PhD at the University of Groningen, in the Netherlands, while also working as an assistant in the university's medical department.

On May 10, 1940, Germany invaded the Netherlands. At the start of the invasion, Kolff was attending a funeral in The Hague. He decided to leave the funeral early and head to the city's main hospital, which was already overwhelmed with casualties, to ask to set up what would be Europe's first blood bank (see chapter 22). The hospital agreed and provided Kolff with a car. Kolff drove through the city collecting tubes, bottles, needles, citrate, and other paraphernalia—all while dodging sniper fire and avoiding falling bombs. Four days later, the blood bank at The Hague's main hospital was operational and saved the lives of hundreds of people.

A month after Germany's invasion, Kolff's Jewish mentor in a Groningen hospital committed suicide and was replaced by a Nazi official. Kolff, not wanting to work with the Nazis, transferred to a small hospital in Kampen for the remainder of the war. Also during the war, in his home, Kolff concealed a Jewish colleague's young son from the Nazis.

Early in Kolff's career as a physician, he witnessed a 22-year-old patient's painful death from kidney failure. At the time, he could do nothing to save the young man, but it struck him that if he had been able to remove the waste that healthy kidneys usually filter away (known as urea), the patient might have lived. Kolff noted, "I realised that removing 22 cubic centimetres of toxicity from his blood would have saved his life." After that traumatic experience, Kolff devoted himself to researching kidney failure.

Kolff developed his first prototype dialyzer machine in 1943. As the Netherlands was still under German occupation, materials were in short supply, but he managed to build his machine using orange juice cans, used auto parts, and cellophane sausage skins wrapped around a cylinder that rested in an enamel bathtub of cleansing fluid. Kolff's machine drew the blood of a patient into a bath, cleaned it, and then passed it back into the patient's body. Over a two-year period, he attempted to treat 15 patients with the machine, but all attempts were unsuccessful. Despite these failures, Kolff persisted.

A breakthrough arrived a month after the war ended in August 1945, when Kolff treated a 65-year-old woman imprisoned for being a Nazi collaborator and in a coma due to renal failure. Many of his fellow countrymen disapproved of treating the woman due to her Nazi ties. However, Kolff persisted in his Hippocratic duty and, after hours of treatment, she awoke and went on to live for another six years before dying of causes unrelated to her kidney problems. A year later, in 1946, Kolff was awarded a PhD from the University of Groningen.

After demonstrating the effectiveness of his artificial kidney, Kolff made dialysis machines and sent them to hospitals all over the world. The machines quickly gained popularity, and in 1948, the artificial kidney was used to perform the first human dialysis in the United States, at Mount Sinai Hospital in New York City.

Kolff immigrated to the United States in 1950 and joined the Cleveland Clinic Foundation. During his time in Cleveland, he helped develop the first heart-lung machines that oxygenated blood and maintained the heart and pulmonary functions of a patient during cardiac surgery. In 1967, Kolff became head of the University of Utah's Division of Artificial Organs and Institute for Biomedical Engineering. While at Utah, he led the medical team that developed the world's first artificial heart, which was successfully implanted in a patient in December 1982.

Although Kolff officially retired in 1986, he continued to work as a research professor and director of the Kolff Laboratory at the University of Utah until 1997. In his lifetime, he was awarded more than 12 honorary doctorates from universities all over the world and more than 120 international awards, including the AMA Scientific Achievement Award in 1982, the Albert Lasker Award for Clinical Medical Research in 2002, and the Russ Prize in 2003. In 1990, *Life* magazine listed Kolff as one of the 100 Most Important Persons of the 20th Century. Kolff died on February 11, 2009, just three days short of his 98th birthday.

Willem Kolff is often called the father of artificial organs, and the technology he created has saved millions of lives around the world.

36

Alan Turing

Father of computer science

Alan Turing was an English mathematician, computer scientist, and cryptanalyst who is best known for his contributions to the field of computer science and for developing a machine that cracked the Nazis' Enigma code during World War II. The Enigma machine was an encryption device that was used extensively by the Nazi forces during World War II to send messages securely. Turing's work in creating a machine that could break the encrypted German messages gave Allied forces a huge advantage during the war. Some historians have estimated that thanks to Turing's work, World War II was shortened by at least two to three years. By cutting the war short, Turing's work likely saved millions of lives and, by cracking the German code, helped ensure an Allied victory.

Alan Turing was born on June 23, 1912, in London. At an early age, he displayed signs of high intelligence and, after enrolling at Sherborne School at the age of 13, he developed a passion for mathematics and science. In 1931, Turing was accepted to study at the University of Cambridge. He graduated three years later with first-class honors in mathematics. The Cambridge faculty were so impressed with his work that he was elected a fellow of King's College, Cambridge, at just 22 years of age.

In 1936, Turing published a seminal paper, "On Computable Numbers, with an Application to the *Entscheidungsproblem* [decision problem]." In that paper, Turing presented the idea of a universal machine—later called the Turing machine—that could solve complex calculations. Many consider Turing's paper a foundational work in the field of computer science and artificial intelligence, as it foreshadowed how a modern digital computer could work.

The same year, Turing moved to New Jersey to study for a PhD in mathematics at Princeton University. He graduated with his PhD in just two years and returned to his fellowship at Cambridge in 1938. A few months later, he was asked to join the Government Code and Cypher School (GCCS), a British code-breaking organization. With the outbreak of World War II in September 1939, following the German invasion of Poland, Turing moved to the GCCS's wartime headquarters at Bletchley Park, Buckinghamshire.

A few weeks before Britain declared war on Germany, the Polish government gave the British government the details of its experts' work on cracking the German Enigma machine. Although Polish intelligence had some success in cracking the

Enigma code at the outbreak of the war, the Nazis increased the machine's security and began to change the cipher daily. That meant that Turing and his team had just 24 hours to crack the Enigma code and translate the messages before the cipher was changed again.

Turing played a key role in creating a machine known as the Bombe, a device that helped significantly reduce the work involved in cracking the Enigma code. By mid-1940, Luftwaffe communications were being unencrypted and read at Bletchley Park.

Once the German Air Force communications had been cracked, Turing turned his attention to decrypting the more complex German naval communications. This work was of vital importance because U-boats were destroying many cargo ships loaded with essential supplies going from North America to Britain. So many supply ships were being destroyed that Prime Minister Winston Churchill's analysts calculated that Britain would soon be starving.

Thankfully, by 1941, Turing personally cracked the different form of Enigma code that was being used by the German U-boats. With the U-boats revealing their positions in their communications with one another, Allied cargo ships could be diverted away from the "wolfpacks" of Nazi submarines. After World War II, Churchill confessed that "the only thing that ever really frightened me during the war was the U-boat peril."

Many historians agree that had Turing not cracked the German naval Enigma code, the Allied invasion of Europe—the D-Day landings—would have likely been delayed by at least a year. Any delay in invading mainland Europe would have enabled the Germans to strengthen their coastal defenses and prolong the time it took the Allied forces to reach Berlin and would have delayed the liberation of the Nazi concentration camps.

After the war ended in 1945, Turing was awarded the Order of the British Empire, or OBE, for his services to the country, and he moved to London to work for the National Physical Laboratory. During his time in London, he led the design work on the Automatic Computing Engine, the world's first stored-program computer. Although the complete version of Turing's design was never built, the adapted concept significantly influenced the design of the English Electric DEUCE and the American Bendix G-15, the world's first personal computers.

In 1952, Turing was prosecuted for homosexual acts after the police discovered that he had been in a sexual relationship with a man. To avoid prison, he agreed to undergo chemical castration through a series of synthetic estrogen injections. As a result of his conviction, Turing's security clearance was revoked, and he was barred from continuing his work with cryptography at GCCS, which had become Government Communications Headquarters, or GCHQ, in 1946.

Despondent at being shunned from the field he had revolutionized, Turing committed suicide in 1954 at age 41. The immense legacy of Turing's life did not fully come to light until the 1970s, when the secret work done at Bletchley Park was declassified.

Turing's impact on computer science is celebrated by the annual Turing Award, the highest accolade in the field of computing. In 1999, *Time* magazine named Turing one of the 100 Most Important People of the 20th Century.

In December 2013, Queen Elizabeth II formally pardoned Turing. In January 2017, the British government enacted Turing's Law, which posthumously pardoned thousands of gay and bisexual men who were convicted under historic legislation that outlawed homosexual acts.

Turing is often considered the father of computer science for his work in conceptualizing the world's first personal computer. If that achievement weren't enough, Turing's contribution to cracking the Enigma code at Bletchley Park is also credited with shortening World War II by several years, which saved millions of lives and helped secure an Allied victory, saving democracy in Europe.

37

Wilson Greatbatch

Creator of the first implantable pacemaker

Wilson Greatbatch was an American engineer and inventor who created the first implantable pacemaker, which uses electrical pulses to ensure that the patient's heart beats at a normal rate. The life expectancy for people with a pacemaker is the same as that for the general population. Receiving a pacemaker is generally considered a low-risk operation, and every year, hundreds of thousands of people are implanted with one. The World Economic Forum estimates that since its invention, the pacemaker has helped save eight million lives.

Wilson Greatbatch was born on September 6, 1919, in Buffalo, New York. His father was an English carpenter who immigrated to the United States in the early 1900s. Greatbatch's American mother died when the future inventor was a young boy. From an early age, Greatbatch had an interest in electronic gadgetry and, as a teenager, he enjoyed assembling radios. After he finished high school in 1936, he put his knowledge of electronics to use by joining the U.S. Navy as a wireless operator and repairer of electronic equipment.

Greatbatch served in both the Atlantic and the Pacific during World War II before being honorably discharged in 1945 as a chief aviation radioman. After spending a year working as a telephone repairman, he enrolled at Cornell University to study electrical engineering. To supplement his income, Greatbatch ran the university's radio transmitter. He also assisted with the electronics that ran the university's radio telescope.

Greatbatch graduated from Cornell in 1950 with a bachelor's in electrical engineering and began studying for his master of science in electrical engineering at the University of Buffalo. In 1952, he became an assistant professor at the electrical engineering department at the University of Buffalo (now called the State University of New York at Buffalo, or University at Buffalo for short).

In the early 1950s, Greatbatch learned about heart block—a condition in which nerves fail to send electrical impulses to the heart, causing irregular heartbeats—when two surgeons visited Cornell University. At the time, the medical procedure to combat heart block was to deliver painful electric shocks using bulky external equipment. Intrigued, Greatbatch started to consider a way to create a smaller, implantable device to help the heart beat regularly.

In 1956, while still working as an assistant professor in Buffalo, Greatbatch made the most important discovery of his lifetime, thanks to a fortuitous error.

While trying to create an instrument that could record heartbeats, Greatbatch accidentally soldered a wrong-size resistor into the circuit. Rather than simply recording electrical pulses that could be used to monitor heartbeats, Greatbatch's mistake caused the device to generate regular pulses of electrical current. Realizing that he had found a way to both electrically simulate and stimulate a heartbeat, Greatbatch later recalled that he "just stared at the thing in disbelief, thinking this was exactly the properties of a pacemaker."

Over the next two years, Greatbatch successfully miniaturized the device down to two cubic inches. After encasing the pacemaker in an epoxy resin to protect it from bodily fluids, he was keen to test it. Thanks to the help of William Chardack, a surgeon at Buffalo's veterinary hospital, the pacemaker, powered by a mercury-zinc battery, was successfully implanted in a dog in May 1958. With this experiment, Greatbatch was able to demonstrate that his device could control the dog's heartbeat.

Later that year, with $2,000 in savings (approximately $21,150 in 2023 dollars), Greatbatch left his job at the University of Buffalo and continued to develop his invention in his garden shed. By 1960, Greatbatch's pacemaker was successfully implanted in the first human patient, a 77-year-old man who went on to live another 18 months. In the same year, nine more patients received the implant.

Greatbatch patented his pacemaker in 1962 and licensed it to Medtronic Inc., a leading manufacturer of medical equipment. However, he soon realized that because his pacemaker could last only two years because of battery constraints, a more reliable source of power was needed to make it a long-term success. In 1972, Greatbatch and his team acquired the rights to the pacemaker and began using newly developed lithium iodine batteries in implantable pacemakers. These new batteries meant pacemakers could last more than 10 years. The lithium iodine battery is still used in pacemakers.

In 1970, Greatbatch founded Wilson Greatbatch Ltd. (now Greatbatch Inc.). By 1972, Greatbatch's new pacemakers, which could last more than 10 years, were on the market and being implanted in thousands of patients across the world. Today, Greatbatch Inc. is a world-leading battery supplier for medical devices and the largest producer of pacemakers in the United States.

Greatbatch received numerous honors in his lifetime. He was given four honorary doctorates, and in 1988, he was inducted into the National Inventors Hall of Fame. In 1990, he was awarded the Lemelson-MIT Lifetime Achievement Award. In 2001, he was granted the highest honor from the National Academy of Engineering, which he shared with Earl Bakken, who invented the external pacemaker. In 1983, Greatbatch's implantable pacemaker was also recognized as one of the two major engineering contributions to society of the previous 50 years by the National Society of Professional Engineers.

Greatbatch and his wife established the Eleanor and Wilson Greatbatch Foundation, which focused on donating money to schools and other educational causes. Greatbatch passed away on September 27, 2011, in Williamsville, New York. At the time of his death, he held over 220 patents. Even late in life, he remained interested in researching everything from nuclear-powered spaceships to solar-powered canoes.

Thanks to the work of Wilson Greatbatch, millions of people around the world have been saved from a painful early death. Every year, hundreds of thousands of people continue to have their lives saved thanks to the implantable pacemaker.

38
Malcom McLean

Businessman who revolutionized shipping

Malcom McLean was an American truck driver and later businessman who developed the modern intermodal shipping container. McLean's development of standardized shipping containers significantly reduced the cost of transporting cargo around the world. Lower shipping costs significantly boosted international trade, which in turn has helped hundreds of millions of people to lift themselves out of poverty. McLean's innovation of "containerization" remains a vital pillar of our interconnected global economy today.

Before McLean developed the standardized shipping container, nearly all of the world's cargo was transported in a diverse assortment of barrels, boxes, bags, crates, and drums. A typical ship in the pre-container era contained as many as 200,000 individual pieces of cargo that had to be loaded onto the ship by hand and unloaded the same way. The time it took to load and unload the cargo often equaled the time that the ship needed to sail between ports. That inefficiency contributed to keeping the cost of shipping very high. This is where McLean enters the story.

Malcolm (later Malcom) McLean was born in November 1913 in Maxton, North Carolina. When he graduated from high school in 1935, his family lacked the funds needed to send him to college. Instead, he began working as a driver for his siblings' trucking company.

In 1937, McLean made a routine delivery of cotton bales to a port in North Carolina for shipment to New Jersey. As he couldn't leave until his cargo had been loaded onto the ship, he sat for hours watching dozens of dock hands load thousands of small packages onto the ship. McLean realized that the current loading process wasted enormous amounts of time and money and began to wonder if there could be a more productive alternative.

In 1952, McLean thought of loading entire trucks onto a ship to be transported from North Carolina to New York. Although this idea would dramatically reduce loading times, he soon realized that these "trailer ships" would be inefficient because of the large amount of wasted cargo space.

Mclean modified his original design so that just the containers, and not the trucks' chassis, were loaded onto the ship. He also developed a way for the containers to be stacked on top of one another. That was the origin of the modern-day shipping container.

In 1956, McLean secured a $22 million bank loan to buy two World War II–era tanker ships and convert them to carry his containers. Later that year, one of his two ships, the SS *Ideal-X*, was loaded with 58 containers and sailed from New Jersey to Houston. At the time, McLean's company offered shipping rates that were 25 percent lower than those of his competitors, as well as the ability to lock the containers. This security measure appealed to many customers, as it eliminated much of the theft and corruption that had long been prevalent at ports.

By 1966, McLean had launched his first transatlantic service, and three years later, he started a transpacific shipping line. As the advantages of McLean's container system became clear, bigger ships, more sophisticated containers, and larger cranes to load cargo were developed.

In 1969, McLean sold his first shipping company for $530 million (about $4.4 billion in 2023 dollars) and went on to start a series of other business ventures. Most notably, he purchased the shipping company United States Lines in 1978 and built a fleet of 4,400 container ships. McLean continued to refine the design of his shipping containers for the rest of his life. He died at the age of 87 in Manhattan in 2001. When he died, *Forbes* magazine called McLean "one of the few men who changed the world."

In 1956, hand-loading loose cargo onto a ship in a U.S. port cost $5.86 per ton ($65.86 in 2023 dollars). However, thanks to McLean's new containers, the price was reduced to just 16 cents per ton ($1.80 in 2023). As bestselling author Matt Ridley has noted, "The development of containerisation in the 1950s made the loading and unloading of ships roughly twenty times as fast and thereby dramatically lowered the cost of trade."

This dramatic reduction in shipping costs boosted international trade. That means that consumers now have access to goods from around the world at a price much lower than was previously thought possible. Similarly, reduced shipping costs have helped to boost the living standards of hundreds of millions of people in export-oriented developing countries over the past few decades.

Without McLean's containers, global trade would be far below what it is today, and we all would be worse off economically.

39

Norman Borlaug

Father of the green revolution

Norman Borlaug was an American agronomist and plant scientist commonly dubbed the father of the green revolution. In the mid-1940s, he began to develop and refine semi-dwarf (shorter plants with stronger stems), high-yield, disease-resistant wheat crops. Borlaug's wheat varieties also grew faster than regular wheat and could withstand harsh climates. By introducing his wheat crops in dozens of nations around the world, Borlaug played a crucial role in ushering in the green revolution, which drastically increased global food production in the second half of the 20th century. It has been estimated that Borlaug's work has helped save more than one billion lives.

Norman Ernest Borlaug was born on March 25, 1914, on his grandparents' farm near Cresco, Iowa. From an early age, he developed a love of agriculture and farming by working on the 106-acre farm. Borlaug obtained his primary education at a small, one-teacher, one-room rural school. At age 13, he began attending Cresco High School, where he excelled in biology and wrestling.

After he initially failed the entrance exam to the University of Minnesota in 1933, Borlaug was accepted at the school's General College. After half a year of study, he transferred to the forestry program at the university's College of Agriculture. To finance his studies, Borlaug took odd jobs and put his education on hold. In 1935, he became a team leader in the Civilian Conservation Corps. This voluntary New Deal government work program supplied manual labor jobs related to conserving and developing natural resources for unemployed young men between 1933 and 1942. As many of the people he worked with were impoverished and hungry, he later recalled, "I saw how food changed them. . . . [A]ll of this left scars on me." Borlaug became troubled by the fact that so many people were suffering from malnutrition and starvation, both in the United States and abroad.

Toward the end of his undergraduate studies in 1937, Borlaug attended a lecture by plant pathologist Elvin Charles Stakman, who was developing methods to identify and combat diseases in wheat crops. In his speech, Stakman explained that he had discovered unique breeding methods that could produce crops resistant to rust, a parasitic fungus that greatly diminished yields of wheat, oats, and barley. Soon after the lecture, Borlaug enrolled in a master of science program in plant pathology under

Stakman at the University of Minnesota, where he earned a PhD in plant pathology and genetics in 1942.

After receiving his doctorate, Borlaug was hired by the DuPont Company as the lead microbiologist studying bactericides and fungicides. However, following the bombing of Pearl Harbor, his lab was converted to conduct research for the U.S. armed forces. In July 1944, despite DuPont offering to double his salary if he stayed, Borlaug moved to Mexico and began working as a head research scientist in charge of wheat improvement at the Rockefeller Foundation.

The aim of Borlaug's task force was to boost wheat production in Mexico, which at the time was a net importer of grain, by teaching local farmers better methods to increase crop productivity. In his memoir, published in 1997, Borlaug stated that his first few years in Mexico were difficult. According to his book, he lacked good equipment and trained scientists, and local farmers were usually hostile to the program because of significant crop losses caused by stem rust in preceding years. Borlaug noted, "[I]t often appeared to me that I had made a dreadful mistake in accepting the position in Mexico."

While in Mexico, Borlaug soon became obsessed with developing better crops that were higher yielding and resistant to rust, pests, and harsh climates. During his first decade in the country, Borlaug and his team experimented with more than 6,000 individual crossbreeds of wheat. Eventually, he came up with a new high-yield wheat crop that was resistant to stem and leaf rust (two of the most devastating diseases plaguing wheat crops in those days), resistant to parasites, and not highly sensitive to the number of daylight hours, which meant it could grow in various climates. Importantly, the new strain was a dwarf variety, meaning it was shorter and sturdier than traditional varieties; it would grow faster and not collapse under heavy loads of grain.

The new wheat variety was an instant success, and wheat yields boomed. The Mexican government soon asked Borlaug to expand the program across the country. By 1963, 95 percent of Mexico's wheat was of Borlaug's variety, and the nation's wheat harvest increased to almost six times what it had been when he first set foot in the country 19 years earlier. As a result, Mexico became a net exporter of wheat by 1963.

Following his success in Mexico, Borlaug became inspired to expand his work to other parts of the world. In March 1963, he visited India with 100 kilograms (220 pounds) of seed from four of his most promising strains, at a time when the country faced a threat of massive food shortages. In 1965, after extensive testing, Borlaug's team ordered about 500 tons of semi-dwarf seed varieties to be sent from Mexico to India and Pakistan.

Unfortunately, Borlaug's seed convoy faced problems from the start. The first shipment was held up in Mexican customs and could not be shipped from the port at Guaymas, on the Gulf of California, in time for planting. To circumvent this problem, Borlaug's team sent the seed in a 30-truck convoy to Los Angeles, where it could be shipped to India. Along its journey, the shipment faced delays at the Mexico-U.S. border and was later stalled by race riots in Los Angeles that obstructed the route to the port. Once the seeds finally reached the port, a Mexican bank would not honor the Pakistani government's payment of $100,000 because the check had three misspelled words.

Eventually, Borlaug's shipment began its voyage to India and Pakistan, but it was far from smooth sailing. Just after the shipment left Los Angeles, Borlaug learned that the Indian government was planning to refuse fertilizer imports because it wanted to build up the country's domestic fertilizer industry. Luckily, that policy was abandoned once Borlaug famously shouted at India's deputy prime minister.

But the problems were just beginning. As Borlaug later noted, "I went to bed thinking the problem was at last solved and woke up to the news that war had broken out between India and Pakistan."

Once the seed arrived in India and Pakistan, Borlaug and his team continued to work tirelessly planting seeds. Some of the fields where they were working were within sight of artillery flashes. If that wasn't enough, a week after the shipment arrived, Borlaug discovered that excessive fumigation at customs had killed half the seeds. In response, he immediately ordered all hands to double the number of seeds they planted.

Despite the holdups and late planning, the yields of Borlaug's crops were higher than any ever harvested in South Asia. In 1965 alone, Indian wheat yields rose by 70 percent compared with the previous year. The proven success of his harvests and the fear of wartime starvation meant that Borlaug got the go-ahead from the Pakistani and Indian governments to roll out his program on a larger scale. The following harvest was even more bountiful, and wartime famine was averted.

Both nations praised Borlaug immensely. The Pakistani agriculture minister took to the radio applauding the new crop varieties, while his Indian counterpart went as far as to plow his cricket pitch with Borlaug's wheat. After a huge shipment of seeds in 1968, the harvest in both countries boomed. The harvest was so bountiful that there were not enough people, carts, trucks, or storage facilities to store all the excess crop.

This extraordinary transformation of Asian agriculture in the 1960s and 1970s nearly banished famine from the entire continent. By 1974, wheat harvests had tripled in India, and for the first time, the subcontinent became a net exporter of the crop. In Pakistan, wheat yields almost doubled between 1965 and 1970, and the nation was self-sufficient in wheat by 1968. Today, food production in India and Pakistan has increased faster than population growth, and both countries produce about seven times more wheat than they did in 1965. The use of high-yield farming in India has prevented approximately 100 million acres, an area roughly the size of California, of wilderness from being converted into farmland.

After the Indo-Pakistani war, Borlaug spent years working to spread his crops in China and later Africa. In 1970, Borlaug was awarded the Nobel Peace Prize for his accomplishments. He retired from his work at the Rockefeller Foundation in 1979 but stayed on as a senior consultant. In 1984, he began teaching and researching at Texas A&M University and became a distinguished professor of international agriculture.

In his later years, he took up several charitable roles. Despite criticism from some other plant scientists, who feared the potentially adverse impact on soil quality and nature, Borlaug was an outspoken advocate for increased agricultural biotechnology and created what agriculture economists call the Borlaug hypothesis, which posits that increasing crop yields is one of humanity's best tools to curb deforestation. In 2000, at the 30th anniversary of his Nobel Prize acceptance, Borlaug expanded on his hypothesis and noted, "Had the global cereal yields of 1950 still prevailed in 1999, we would have needed nearly 1.8 billion ha [hectares] of additional land of the same quality—instead of the 600 million that was used." Borlaug was also an advocate for using DDT as an insecticide, something pioneered by fellow hero of progress Paul Hermann Müller (see chapter 31).

For his work, Borlaug received dozens of awards and honors. He is one of only seven people to have received both the Congressional Gold Medal and the Presidential Medal of Freedom—in addition to the Nobel Peace Prize. Borlaug was awarded about 50 honorary doctorates and numerous fellowships, and several buildings and schools are named after him. It is said that he was particularly satisfied when the people of Sonora, Mexico, where he did some of his first experiments, named a street after him.

Borlaug died of lymphoma on September 12, 2009, at 95 years old, at his home in Dallas.

Borlaug's wheat helped to save hundreds of millions of people, if not over a billion, from dying of starvation. Borlaug's wheat varieties are regularly credited for ushering in the green revolution, which drastically increased global food production and greatly aided environmental protection by helping prevent hundreds of millions of acres of wilderness from being turned into farmland.

40

Jonas Salk

Polio vaccine pioneer

Jonas Salk pioneered the world's first effective polio vaccine. Polio is a highly infectious viral disease that is most often transmitted by drinking water that has been contaminated with the feces of someone carrying the virus. The virus spreads easily in regions with poor sanitation. The symptoms include fever, fatigue, headache, vomiting, stiffness, and pain in the limbs. Most infected patients recover. In 1 of 200 cases, the virus attacks the nervous system, leading to irreversible paralysis. Of those paralyzed, between 5 and 10 percent die when their breathing muscles become immobilized.

Polio has a relatively long incubation period. It can spread for several months before being detected, which makes outbreaks extremely difficult to monitor. According to Max Roser of the University of Oxford, "Up to the 19th century, populations experienced only relatively small outbreaks [of polio]. This changed around the beginning of the 20th century. Major epidemics occurred in Norway and Sweden around 1905 and later also in the United States."

The first major outbreak of polio happened in the United States in 1916, when the disease infected 27,000 people and killed more than 7,000 people. The second major outbreak in 20th-century America happened in the 1950s. It is here that Jonas Salk enters the story.

Jonas Edward Salk was born on October 28, 1914, in New York City. He became passionate about biochemistry and bacteriology during his time at the New York University School of Medicine. After he graduated in 1939, he started working at the prestigious Mount Sinai Hospital. Salk's focus shifted to researching polio vaccinations in 1948, when he was recruited to work at the National Foundation for Infantile Paralysis, which President Franklin D. Roosevelt, himself a polio sufferer, had helped to set up.

After a large polio outbreak across the United States in 1952, donations began pouring into the foundation, and in the spring of 1953, Salk put forward a promising anti-polio vaccine. The foundation quickly began trials on 1.83 million children across the United States. These children became known as the polio pioneers. Salk's foundation received donations from two-thirds of the American population, and a poll even suggested that more Americans knew about these field trials than knew the full name of President Dwight David Eisenhower.

On April 12, 1955, Salk's supervisor, Thomas Francis, announced that Salk's vaccine was safe and effective in preventing polio. Just two hours later, the U.S. Public Health Service issued a production license for the vaccine, and a national immunization program began.

Shortly thereafter, Dr. Albert Sabin, a Polish American medical researcher working at the National Institutes of Health, introduced a polio vaccine that could be administered orally, which made vaccination efforts less expensive because trained health workers weren't needed to administer injections. From a record 58,000 cases in 1952, the United States was declared polio-free in 1979.

In 1988, the World Health Organization founded the Global Polio Eradication Initiative (GPEI) to administer the vaccine around the world. When the GPEI began its efforts, polio paralyzed 10 children for life every 15 minutes across 125 countries. Since 1988, more than 2.5 billion children have been immunized, and incidents of polio infections have decreased by more than 99.99 percent—falling from 350,000 annual cases to just 22 new cases across three countries in 2017. In 2020, the Africa Regional Certification Commission, an independent body, declared Africa free from wild polio; unfortunately, a few sporadic cases resurfaced in Malawi in late 2022.

Following his discovery of the vaccine, Salk received dozens of awards, the Presidential Medal of Freedom, four honorary degrees, half a dozen foreign decorations, and letters from thousands of thankful fellow citizens. In 1963, Salk established the Jonas Salk Institute for Biological Studies, a world-class research facility that focuses on molecular biology and genetics, neurosciences, and plant biology. Salk devoted his later years to researching a vaccine for HIV and AIDS. He died of heart failure on June 23, 1995, in San Diego.

Salk's work has saved hundreds of millions of people from crippling paralysis and millions from death. Thanks to his vaccine, a disease that has plagued humanity since the time of ancient Egypt has been almost completely eradicated, and within a few years, the disease will hopefully be consigned to history.

41
Benjamin Rubin

Inventor of the bifurcated needle

Benjamin Rubin was an American microbiologist who invented the bifurcated needle, a valuable tool against smallpox. Rubin's needle was instrumental in the World Health Organization's (WHO's) 1980 campaign that led to the full eradication of smallpox, the only infectious disease for which that goal has been attained. In the mid-1970s, Rubin's bifurcated needle was used to administer more than 200 million vaccinations annually. The World Economic Forum estimates that the bifurcated needle has saved more than 100 million lives and prevented hundreds of millions more people from contracting smallpox.

Benjamin Rubin was born on September 27, 1917, in New York City. As a child, he was fascinated by science. In 1934, Rubin enrolled at City College of New York to study biochemistry. He received a bachelor of science in 1937. In 1938, Rubin was awarded a master of science in biology from Virginia Tech. After several laboratory jobs, he relocated to Yale University in 1944 to work as a research assistant and study for his PhD.

In 1947, Rubin was awarded a doctorate in microbiology from Yale University and went on to work at several laboratories and universities. In 1954, he became a professor of public health and preventative medicine at Baylor University. In 1960, Rubin took a job at Wyeth Laboratories in Pennsylvania, where he created his world-changing invention.

During the 1960s, smallpox killed more than two million people every year. With no cure available, if a person became infected with smallpox, nothing could be done to treat the disease. Before Rubin's bifurcated needle, specially adapted jet injectors (gun-like devices) were used for mass vaccination efforts. However, the jet injectors were expensive to maintain, required specialist training to use, and were often unreliable. That meant that hundreds of millions of people living in predominantly poorer and sparsely populated places remained vulnerable to catching smallpox.

Rubin created the bifurcated needle by working on the eyelet of a sewing machine needle. He ground it down until it became fork-shaped (with two prongs). The bifurcated needle is a narrow steel rod approximately 2.5 inches (6 centimeters) long, with two prongs at one end. When the bifurcated end of the needle is dipped into a vial of freeze-dried smallpox vaccine, the correct amount of vaccine is contained between the two prongs. The needle is then used to puncture a patient's upper arm 15 times in a small circular area.

Unlike the jet injector, Rubin's needle was very cheap to produce: 1,000 needles cost less than five dollars. The bifurcated needle used substantially less serum and could be continually re-used after being sterilized with boiling water or passed through a flame. Moreover, the patient's skin did not need to be disinfected beforehand.

While a jet injector could deliver about 25 vaccines per vial of smallpox vaccine, the use of the bifurcated needle yielded more than 100 doses per vial. Furthermore, the technique to vaccinate someone using the bifurcated needle could be learned by anyone in a few minutes without having to rely on a trained medical professional. All these advantages led the bifurcated needle to be quickly adopted as a cost-effective alternative to the clunky, expensive, and often unreliable jet injectors.

The bifurcated needle was the primary tool in WHO's smallpox eradication campaign between 1966 and 1977. During the campaign's final years, the bifurcated needle was used to deliver more than 200 million vaccinations per year. In 1980, WHO declared that smallpox had been defeated, meaning that for the first time in human history, humanity had successfully eradicated a deadly disease. The World Economic Forum has estimated that the bifurcated needle has saved more than 130 million lives since its invention in 1961.

In 1984, Rubin became a professor at the Philadelphia College for Osteopathic Medicine. After contributing to more than 150 science journals throughout his career, he retired in 1995. Rubin received numerous academic awards in his lifetime. In 1992, he was inducted into the Inventors Hall of Fame. Rubin died on March 10, 2010, at the age of 93.

Thanks in part to the work of Benjamin Rubin, smallpox, a disease that had haunted humankind for millennia, has been obliterated. Rubin helped save the lives of more than 100 million people and radically changed the world for the better.

42

James Elam and Peter Safar

Developers of CPR

James Elam and Peter Safar were physicians who developed and popularized modern cardiopulmonary resuscitation, better known as CPR. Elam and Safar's modern method of CPR is taught to people across the world as the go-to method to revive a nonbreathing person. The World Economic Forum has estimated that Elam and Safar's technique has already saved five million people and continues to save hundreds of thousands of people every year.

Before CPR, people used many peculiar methods to bring unresponsive patients around. In the medieval era, flagellation was used. In the 16th century, people commonly used bellows from a fireplace to blow hot air and smoke into the patient's mouth. In the late 19th century, stretching an unconscious person's tongue while tickling their throat with a feather was recommended. As late as the 1950s, people were advised to revive patients by lifting the unconscious person's arms and applying pressure to their chest. These ineffective resuscitative methods have largely been abandoned thanks to the work of Elam and Safar.

James Elam was born on May 31, 1918, in Austin, Texas. He was born prematurely, weighing just two pounds at birth, and had trouble breathing as a child. He later declared that his childhood experiences with breathing difficulties helped influence his future career decisions. In 1942, Elam graduated with a bachelor of arts from the University of Texas. Three years later, he earned an MD from Johns Hopkins School of Medicine.

After an internship at the U.S. Naval Hospital in Bethesda, Maryland, in 1946 Elam decided to pursue further training in physiology at the University of Minnesota. During his time there, he became interested in respiration. After hearing about how midwives in Europe had had some success using a mouth-to-mouth technique on newborns, he began to wonder whether there was a more effective way to resuscitate patients than the existing pressure and arm-lifting technique.

After a couple of years studying surgery at Barnes Hospital in St. Louis, Missouri, Elam began to focus on respiratory physiology. Although he was not the first person to create a ventilator, he designed several of his own, and they proved capable of providing respiratory assistance to hospitalized patients. Elam knew that his ventilators would be useless for patients who experienced respiratory difficulties outside a hospital, however, so he began working on what would come to be known as rescue breathing.

In 1954, Elam successfully demonstrated that exhaled air could provide adequate oxygen for resuscitation. He tried to bring his findings to the attention of the medical community and general public, but it wasn't until he received help from Peter Safar that Elam's discovery became common knowledge.

Peter Safar was born on April 12, 1924, in Vienna, Austria. He began medical school at the age of 19 at the University of Vienna. Safar graduated with an MD in 1948. In 1949, he accepted a surgical fellowship at Yale University.

After a brief time training in anesthesiology at the University of Pennsylvania, Safar moved to Lima, Peru, where he founded the country's first academic anesthesiology department. In 1954, he became chief of the anesthesiology department at Johns Hopkins Hospital in Baltimore. Two years later, Safar met Elam at the American Society of Anesthesiologists' meeting in Kansas City, Missouri.

While sharing a ride back to Baltimore, the pair debated the pros and cons of various resuscitation methods. After Elam explained that he had successfully proved that providing exhaled air to the patient was adequate for resuscitation, Safar suggested and subsequently conducted a series of experiments to prove that mouth-to-mouth breathing of exhaled air could maintain satisfactory oxygen levels in a non-breathing patient.

The pair perfected the initial steps of CPR, which included the head-tilt maneuver to open the patient's airway and the precise method of mouth-to-mouth breathing. These experiments highlighted the failure of the commonly used chest pressure and arm-lifting rescue methods and documented the superiority of Elam's exhaled-air technique.

In 1957, Safar wrote a book titled *ABC of Resuscitation* that described the A (airway) and B (breathing) steps of CPR along with step C (chest compressions). The book became the basis for CPR training. The same year, the U.S. military endorsed Elam and Safar's method of resuscitation. In 1958, Safar published the findings from his CPR experiments in the *Journal of the American Medical Association*.

In 1959, Elam wrote a shorter instructional booklet highlighting his CPR method titled *Rescue Breathing*. The success of the publication led Elam to produce films demonstrating the technique. The two doctors also contributed to the development of the CPR practice mannequin called Resusci Anne.

The compelling evidence for Elam's and Safar's method of CPR meant that the change from the old chest pressure and arm-lifting method to mouth-to-mouth resuscitation was, as Safar described it, "extremely rapid." By 1960, their CPR method had been adopted by the American Red Cross and the National Academy of Sciences. Before long, it was being taught around the world.

In recognition of their work, Elam and Safar received a plethora of awards and honors. In 1960, Elam was awarded a Certificate of Achievement by the U.S. Army. In 1962, the Medical Society of the State of New York awarded Elam its highest honor, the Albert O. Bernstein Award. Safar was awarded Austria's highest civilian honor, the Cross of Honor, in 1999.

On three occasions, he was nominated for the Nobel Prize in Medicine.

Later in life, Elam founded the Society for Obstetric Anesthesia and Perinatology. Until his death on July 10, 1995, he continued to experiment with and modify a variety of anesthetic devices.

Safar went on to create the first academic anesthesiology department and the world's first intensive-care medicine training program at the University of Pittsburgh. He noted that his lifelong goal was to "save the hearts and brains of those too young to die." Safar continued his research until his death on August 3, 2003.

Thanks to the work of James Elam and Peter Safar, emergency medicine has changed for the better. The duo's modern CPR method continues to be taught worldwide and is credited with saving hundreds of thousands of lives every year.

43

Maurice Hilleman

Prolific vaccine creator

Maurice Hilleman was an American microbiologist who developed more than 40 lifesaving vaccines. He developed 8 of the 14 vaccines recommended in the current schedules, including those for mumps, rubella, hepatitis A and B, meningitis, and pneumonia. Hilleman is credited with saving more lives than any other medical scientist in the 20th century.

Hilleman was born on August 30, 1919, on a farm near Miles City, Montana. His mother died two days after he was born. As his father faced the prospect of raising eight children alone, his childless aunt and uncle agreed to raise Maurice on their nearby chicken farm. Hilleman attributed much of his later success to his work on the farm as a boy; since the

1930s, fertile chicken eggs have been used to grow viruses for vaccines.

Due to a lack of funds, Hilleman almost didn't make it to college until his eldest brother loaned him the money to pay tuition. Hilleman graduated first in his class from Montana State University in 1941. After graduating with a bachelor's degree in microbiology and chemistry, he won a fellowship for postgraduate study in microbiology at the University of Chicago. He received a doctorate in 1944.

Hilleman, despite his professors' advising him to pursue a career in academia, wanted to work in the private sector. He argued that academic institutions lacked the resources to move scientific breakthroughs forward and on to market. Upon graduation, he joined the pharmaceutical company ER Squibb & Sons (now part of Bristol Myers Squibb), which ran a virology lab in New Jersey.

Soon after he started working in the laboratory, Hilleman successfully developed a vaccine for Japanese B encephalitis. This infection, which is native to Asia and the western Pacific, had begun to spread to American troops fighting in the Pacific during World War II.

In 1948, Hilleman began working as the chief of the Department of Respiratory Diseases at the Army Medical Center in Silver Spring, Maryland. During his time there, he discovered the genetic changes that take place when the influenza virus mutates and postulated that this mutation would mean that yearly influenza vaccinations would be required to curb infection.

In 1957, after joining Merck & Co. (one of the world's largest pharmaceutical companies) to head its virus and cell biology department, Hilleman discovered the first signs of an impending flu pandemic that was spreading in Hong Kong. Hilleman and his colleagues raced to produce a vaccine and oversaw the production of more than 40 million doses that were immediately distributed across the United States.

Although 69,000 Americans died after catching the virus, the pandemic could have caused millions of deaths if not for Hilleman's efforts. In recognition of his work, the U.S. Army awarded Hilleman the Distinguished Service Medal in 1957.

In 1963, while Hilleman was working at Merck, his daughter, Jeryl Lynn, became ill with the mumps. Hilleman quickly drove to his lab to pick up the necessary equipment to cultivate material from his daughter's infection.

In 1967, the original sample taken from his daughter's throat became the basis for the newly approved mumps vaccine that Hilleman created. It came to be known as the Jeryl Lynn strain. Looking back at the process used to develop the mumps vaccine, a colleague of Hilleman's remarked, "Today's regulation would preclude that from happening."

Hilleman later combined his mumps vaccine with the measles and rubella vaccines, which he had also developed to significantly reduce their side effects, to create the MMR vaccine, the first vaccine ever approved that incorporated multiple live virus strains.

Apart from the vaccines mentioned above, Hilleman also developed vaccines for hepatitis A, hepatitis B, chickenpox, meningitis, pneumonia, and *Haemophilus* influenza type B. In addition, he helped isolate the cold-producing adenoviruses, the hepatitis viruses, and the cancer-causing simian virus 40.

American physician and liver transplant pioneer Thomas Starzl once noted that "controlling the hepatitis B virus scourge ranks as one of the most outstanding contributions to human health of the 20th century" and that the vaccine removed "one of the most important obstacles" in the organ transplant field.

In 1984, at the mandatory retirement age of 65, Hilleman stepped down as senior vice president of Merck Research Labs. Not satisfied with retirement, within a few months he began directing the newly created Merck Institute for Vaccinology. Hilleman continued to work at the Institute for Vaccinology until his death on April 11, 2005, at the age of 85.

Over the years, Hilleman received a series of awards. President Ronald Reagan presented him with the National Medal of Science, America's highest scientific honor, in 1988. He received several medals from governments and organizations around the world for his contributions to public health and received a special lifetime achievement award from the World Health Organization.

Maurice Hilleman is often described as the most successful vaccinologist in history. Many people alive today likely owe their lives to him.

44

Charles Dotter and Andreas Grüntzig

Pioneers of angioplasty

Charles Dotter and Andreas Grüntzig were radiologists who pioneered angioplasty, a surgical procedure for widening narrow or blocked blood vessels. If left untreated, arterial atherosclerosis—plaque buildup in the veins—can cause severe medical problems, including coronary heart disease, heart attacks, strokes, and other maladies. The World Economic Forum estimates that more than 15 million lives have been saved since Dotter's initial discovery and Grüntzig's subsequent improvement of angioplasty.

Charles Dotter was born on June 14, 1920, in Boston, Massachusetts. He attended grammar school and high school in Freeport, Long Island. He was described as a bright and inquisitive child. From an early age, Dotter was interested in mechanical objects and derived great satisfaction from dismantling and rebuilding machines. In 1941, he was awarded a bachelor of arts from Duke University. In the same year, he enrolled in Cornell University's medical school.

After medical school, Dotter completed an internship at the U.S. Naval Hospital in New York City and later did his residency at New York University. At just 30 years old, he was offered a faculty position at Cornell Medical School. Two years later, he became a professor and the chairman of the Department of Radiology at the University of Oregon—the youngest-ever chairman of a radiology department at a major American medical school.

During his time at the University of Oregon, Dotter directed several successful projects. He is considered the founder of a new medical specialty called interventional radiology—a group of techniques that uses x-rays, magnetic resonance imaging (MRI), and ultrasound imaging to guide medical therapies into internal structures of the body.

One of the machines that Dotter created to help in the development of this new field was the x-ray roll-film magazine, which he produced in 1950. For radiologists, the ideal way to visualize a patient's blood flow is with continuous x-ray imaging, a process known as real-time fluoroscopy. Before Dotter's machine, radiographic images had to be made one at a time, and a technician had to manually change x-ray cassettes for each new image. That led to substantial gaps in x-ray imaging. In contrast, Dotter's new x-ray roll-film magazine could produce an image every two seconds.

A large part of Dotter's work involved conducting imaging studies of patients for surgeons. Like all radiologists at the time, he would insert a catheter (a soft hollow tube) into a patient's artery, squirt a dye into the tube, and then take an x-ray to analyze the patient's circulation and check for any potential blockages. Surgeons would then know where to operate on the basis of the x-rays taken by the radiologist.

However, Dotter theorized that instead of using the catheter to merely inject dye into the blocked artery, he could push the catheter through the blockage itself, thereby opening up the blocked artery and improving blood circulation without the need for intrusive surgeries and general anesthesia.

In 1964, Dotter had the opportunity to test his theory after an 82-year-old patient was admitted to the University of Oregon Hospital with pain in her left foot. Dotter found that the patient had a blockage in the superficial femoral artery and that the lack of blood circulation had caused a nonhealing ulcer and gangrenous toes. All the hospital's physicians had recommended leg amputation, but the patient refused. The surgeon in charge of the case suggested that Dotter try his new technique.

On January 16, 1964, Dotter went ahead with his procedure. He began by sliding a series of progressively larger catheters through the blocked artery to slowly dilate the blockage. He then added a stent, which is a small metal mesh tube to prevent the artery from closing again. The procedure was a success, and within minutes the patient's leg had warmed up. A week later, the patient's pain had disappeared, the ulcer soon healed, and she regained full mobility. Despite his initial success, Dotter's ideas were largely rejected by the vascular surgical community. This was until Andreas Grüntzig began using Dotter's methods.

Andreas Grüntzig was born on June 25, 1939, in Dresden, Germany. In 1951, he enrolled at the Thomasschule high school, the oldest public school in Germany. After he graduated with honors in 1957, Grüntzig fled to Heidelberg in West Germany just before the communists closed the East German border.

Grüntzig began studying medicine at Heidelberg University in the fall of 1958. He graduated six years later. For the next five years, he traveled extensively and completed a series of internships across West Germany and the United Kingdom. In the late 1960s, Grüntzig first learned of Dotter's angioplasty procedure at a lecture that Dotter gave in Frankfurt. Inspired by Dotter's efforts, he started to work on different angioplasty techniques. After encountering bureaucratic resistance in West Germany, he decided to move to Switzerland. In 1969, he started to work in the Department of Angioplasty at the University Hospital of Zürich.

In 1971, Grüntzig began using Dotter's angioplasty technique to treat patients. He also began toying with the idea of adding a balloon to Dotter's catheters, which could then be used to expand blocked arteries. Without any funding, Grüntzig worked relentlessly during evenings and weekends to develop his idea of small balloons that were sturdy enough to inflate arteries. Within two years, he had succeeded in creating handmade balloon catheters.

At the American Heart Association (AHA) meeting in 1976, Grüntzig presented his findings and the successes he had had using balloon catheters on animals. Although most attendees were skeptical of his work, Dr. Richard Myler of Saint Mary's Hospital in San Francisco suggested that they collaborate to perform the first human coronary angioplasty using a balloon catheter.

On September 16, 1977, Grüntzig and Myler used Grüntzig's balloon catheter for the first time on an awake human patient. Grüntzig's balloon technique was successful. Moreover, it was both faster and safer than Dotter's previous method of slowly sliding progressively larger catheters through a blocked artery. A year later, when Grüntzig presented the results of his first four balloon-catheter angioplasty cases to the AHA, the audience gave him a standing ovation, and subsequently, balloon angioplasty was quickly accepted throughout the medical community.

In 1978, Dotter and Grüntzig were both nominated for the Nobel Prize for Physiology or Medicine for their pioneering work. Dotter, commonly known as the father of interventional radiology, stayed at the University of Oregon for 33 years, from his arrival in 1952 until his death on February 15, 1985. The University of Oregon named the Dotter Interventional Institute in his honor. Grüntzig immigrated to the United States in 1980 and became director of interventional cardiovascular medicine at Emory University in Atlanta. On October 27, 1985, Grüntzig and his wife died after a plane that he was piloting crashed. Emory University's Andreas Grüntzig Cardiovascular Center was named for him.

Ultimately, Charles Dotter and Andreas Grüntzig created an entirely new field of medical study and pioneered a procedure that has saved more than 15 million lives and prevented millions of amputations.

45

Vasili Arkhipov

Preventer of nuclear war

Vasili Arkhipov was a Soviet naval officer who refused to authorize a Soviet nuclear strike on an American aircraft carrier during the 1962 Cuban Missile Crisis. Arkhipov's actions likely prevented an all-out nuclear war, the consequences of which would have included the deaths of millions, if not billions, of innocent people, a collapse of many nation states and their economies, and an enormous amount of environmental damage. Aptly, the U.S. National Security Archive has declared that Arkhipov "saved the world."

Vasili Arkhipov was born on January 30, 1926, to a peasant family in Staraya Kupavna, a small town on the outskirts of Moscow. After a typical public school education, he enrolled at the Pacific Higher Naval School, a Soviet naval officer training facility, in 1942. Arkhipov first saw military action during the Soviet-Japanese War in August 1945, when he served aboard a minesweeper (the Soviet Union declared war on Japan as World War II neared its end). In 1947, he graduated from naval school and went on to serve on submarines in the Black Sea and the Baltic.

In 1961, Arkhipov was appointed the executive officer of the Soviet Union's new nuclear ballistic missile submarine, the K-19. During the submarine's maiden voyage, its nuclear cooling system developed a leak that threatened to cause the nuclear reactor to melt down. In the face of a potential mutiny, Arkhipov backed the captain and ordered the engineering crew to develop a technical solution to avoid a nuclear meltdown. The crew members were forced to build an emergency cooling system on the fly. The solution required many of the men to work while exposed to high levels of radiation for extended periods, and although the engineers managed to save the ship and prevent a meltdown, the entire crew, including Arkhipov, were irradiated. Because of the exposure to high levels of radiation, all the members of the engineering crew died within a month. Yet even that harrowing event pales in comparison to what Arkhipov experienced the following year.

On October 1, 1962, Arkhipov was made commodore of a flotilla of four submarines that had been ordered to travel from Russia to Cuba. Arkhipov was also appointed sub-commander of the B-59 attack submarine on which he was traveling. The B-59 carried 22 torpedoes, one of which was nuclear and possessed roughly the same destructive power as the nuclear bomb that the United States had dropped on Hiroshima in 1945. Unbeknownst to the crews of the four submarines, the United States had imposed a naval blockade of Cuba on October 4 and had told the Soviets that U.S. Navy destroyers would drop depth charges (explosive warning shots) on any Soviet submarine in Cuban waters to force the vessels to surface. Due to a lack of radio communications, Moscow was unable to relay that information to Arkhipov's crew.

On October 27, a group of 11 U.S. destroyers and one aircraft carrier, the USS *Randolph,* located Arkhipov's submarine off the coast of Cuba and began pummeling it with signaling depth charges. Arkhipov's submarine was too deep underwater to receive any radio traffic, and with each depth charge causing the submarine to shake uncontrollably, those on board did not know whether a war had broken out. On board the submarine, the air-conditioning system had broken and temperatures in some sections reached over 122 degrees Fahrenheit (50 degrees Celsius). The air supply recirculation system worked poorly, and the rising levels of carbon dioxide caused many of the weary crew, who had already been traveling aboard the submarine for almost four weeks, to faint from overheating.

During that stressful situation, the submarine's captain, Valentin Savitsky, believed that the American navy was firing bombs on their vessel and decided that war between the two countries had already broken out. Savitsky ordered the nuclear-tipped torpedo to be readied and aimed at the USS *Randolph.* The political officer on the B-59, Ivan Maslennikov, agreed with the captain's decision. Usually, Soviet submarines armed with nuclear weapons required the permission of only the captain and the political officer to launch their nuclear torpedoes. However, although Arkhipov was only second in command on the B-59, he was the commodore of the entire submarine flotilla, which meant the captain also needed Arkhipov's approval.

Arkhipov refused to approve the launch of the nuclear torpedo and an intense argument broke out among the three officers. Later Soviet intelligence reports quote the captain as saying, "We're gonna blast them now! We will die, but we will sink them all. We will not disgrace our navy."

However, Arkhipov refused to budge and argued that, as no orders had come from Moscow, such extreme measures would be ill-advised. Instead, he advised that the submarine should surface and contact the naval headquarters. Arkhipov was eventually successful in persuading the captain and, as the submarine rose to the surface, it was met by a U.S. destroyer, which ordered it to immediately return to the Soviet Union.

The American forces did not board the submarine or undertake any inspection, so they were not aware that it was armed with a nuclear torpedo. The U.S. Navy, and indeed the wider public, found out about the B-59's nuclear capabilities and the full story of Arkhipov's actions only in 2002, when the former antagonists met in Cuba for the 40th anniversary of the crisis. When discussing the Cuban Missile Crisis, Arthur Schlesinger, an American historian and former adviser to President John F. Kennedy, said, "This was not only the most dangerous moment of the Cold War. It was the most dangerous moment in human history."

Upon their return to Russia, the crew of the submarine were met with criticism from their superiors, as some officers viewed the act of surfacing as one of surrender. One admiral told Arkhipov, "it would have been better if you'd gone down with your ship." After the events of October 1962, Arkhipov continued his navy service. He was promoted to rear admiral in 1975 and became head of the Kirov Naval Academy. In 1982, he was promoted to vice admiral and retired a few years later. Arkhipov settled in a small town near Moscow and died on August 19, 1998, of kidney cancer that may have been caused by the radiation to which he was exposed while aboard the K-19 in 1961.

Had Arkhipov not been on that B-59 submarine in October 1962 or had he given in to pressure from the other officers, the submarine's nuclear torpedo would have vaporized the USS *Randolph.* That, says Russian archivist Svetlana Savranskaya, would have started "a chain of inadvertent developments, which could have led to catastrophic consequences." According to plans laid out by both the Soviet Union and the United States, the likely first targets of a nuclear war would have been Moscow, London, airbases across the United Kingdom, and troop concentrations in Germany. The next wave of bombs would have wiped out "economic targets"—that is, civilian populations—across the world.

Arkhipov received little recognition during his lifetime, but to his wife Olga, Vasili was always a hero. In a 2012 PBS documentary titled *The Man Who Saved the World,* Olga Arkhipova said, "The man who prevented a nuclear war was a Russian submariner. His name was Vasili Arkhipov. I was proud and I am proud of my husband, always." Thanks to Arkhipov, nuclear war was averted, and countless lives were saved.

46

Raymond Damadian, Paul Lauterbur, and Peter Mansfield

Creators of the MRI machine

Raymond Damadian, Paul Lauterbur, and Sir Peter Mansfield were scientists who created and refined the magnetic resonance imaging (MRI) machine. Damadian created the world's first MRI scanner after he realized that cancerous cells produced magnetic resonance signals different from those of normal, non-cancerous cells. Prompted by Damadian's discoveries, Lauterbur developed a way for MRI machines to visualize these cells' signal differences and produce a clear image of the inside of a patient's body. Finally, Mansfield created a technique for MRI scans to be conducted in seconds rather than hours, and for the image that the scanners produced to be significantly clearer and, therefore, more accurate. Each year, hundreds of millions of MRI scans take place. Thanks to their use, untold millions of lives have been extended or saved.

Raymond Damadian was born on March 16, 1936, in New York City to a family of Armenian immigrants. At the age of 10, his interest in detecting cancer was sparked after his maternal grandmother died of breast cancer. As a gifted violinist, Damadian won a scholarship to the University of Wisconsin when he was just 16 years old. While in Wisconsin, Damadian soon realized that his prospects of becoming a successful violinist were slim. Instead, he began to pursue his other passions: math and chemistry.

In 1956, Damadian graduated from the University of Wisconsin-Madison with a degree in mathematics. With the goal of finding better treatments for cancer, he studied medicine at the Albert Einstein College of Medicine in New York. In 1960, Damadian graduated with an MD and enrolled in postgraduate fellowships at Washington University School of Medicine in St. Louis, Missouri, and then at Harvard Medical School in Massachusetts. During this time, Damadian became interested in the field of medical imaging and magnetic resonance. He had experienced severe abdominal pain, and his doctors, who were using conventional x-rays, were unable to discover the cause of his ailment. This led him to seek a better way to examine the body's inner workings.

As a medical student, Damadian had been automatically deferred from the draft during the Vietnam War. However, in the mid-1960s, as American participation in the war escalated, he received orders from the U.S. Air Force to begin active duty. Damadian was stationed at Brooks Air Force Base in San

Antonio, Texas. During his time there, his commanding officers allowed him to continue his personal work on magnetic resonance, provided that he also did some research for the Air Force on the rocket fuel hydrazine. In 1967, Damadian left the military and joined the faculty of the State University of New York Downstate Medical Center in Brooklyn to continue his work on magnetic resonance.

Magnetic resonance functions by exposing atomic nuclei to a magnetic field and radio waves, which then emit other radio waves at consistent frequencies. When radio waves are pulsed through something that is being scanned, the protons in that object or person are stimulated and spin out of equilibrium. When the field is turned off, protons in whatever is being scanned return to their normal spin and produce a radio signal, which can then be measured by receivers in the scanner. Damadian knew that cancerous cells hold more water, and therefore more hydrogen, than healthy cells. In 1969, he theorized that when magnetic resonance equipment scans a body, radio waves would take longer to pass through cancerous tissue than through healthy tissue. This lag could then be used to detect damaged areas.

A year later, Damadian began testing his theory by using magnetic resonance to scan cancerous liver samples from laboratory rats. His experiments were successful. In 1971, he published his findings in the journal *Science*. In the article, he reasoned that cancerous tissues could be detected externally in humans without using radiation, if a large enough scanner was built. This discovery laid the foundation for the basis of the MRI machines we have today. However, Damadian had no way to generate pictures or to clearly visualize the results of his scans. Thankfully, Paul Lauterbur helped change this.

Paul Lauterbur was born on May 6, 1929, in Sydney, Ohio. Fascinated with science as a child, he built his own laboratory in the basement of his parents' house when he was a teenager. After graduating from high school in 1947, he enrolled at the Case Institute of Technology (now Case Western Reserve University) in Cleveland to study chemistry. After graduating with a bachelor of science in 1951, Lauterbur went to work as a research associate at the Mellon Institute (a predecessor to Carnegie Mellon University) in Pittsburgh, Pennsylvania. In 1953, Lauterbur was drafted into the Korean War and worked at the Army Chemical Center in Edgewood, Maryland.

As was the case with Damadian, Lauterbur's superiors allowed him to work on an early magnetic resonance machine. By the time he left the army in 1955, he had published four scientific papers on magnetic resonance. After a two-year tour of duty in the military, Lauterbur returned to the Mellon Institute and enrolled in graduate classes at the University of Pittsburgh. In 1962, he graduated from Pitt with a PhD in chemistry and accepted a position as an associate professor at Stony Brook University in Long Island, New York.

In 1971, after reading Damadian's article in *Science*, Lauterbur became interested in the potential biological uses of magnetic resonance technology. He was disappointed that Damadian's experiments had been done on dead tissue and wondered if there was a way for living tissue to be imaged. Lauterbur knew that Damadian had used a uniform magnetic field. If a nonuniform field were used, he theorized, a clear image of the scan could be created. By adding gradients to the scanner's magnetic field, the MRI machine could determine the origin of the emitted radio waves from what was being scanned, and an image could then be generated.

In 1973, Lauterbur was successful in producing the first ever magnetic resonance image of water in a test tube. After publishing his findings in the journal *Nature*, he soon created the first such image of a living subject: a small clam.

In 1974, Damadian received the first patent in the field of MRI when his 1972 application for the concept of using magnetic resonance to detect cancer was approved. With the help of several graduate students, Damadian eventually built the first human MRI scanner, nicknamed the Indomitable. On July 3, 1977, almost five years after he began testing the machine, the Indomitable achieved the first human MRI scan of one of Damadian's graduate students. The crude two-dimensional image showed the student's heart and lungs.

On the other side of the Atlantic, another scientist, Peter Mansfield, began working on a method to significantly speed up the time it took for MRI machines to complete a scan. Mansfield was born on October 9, 1933, in London. At the age of 15, he expressed an interest in science. Because of his unexceptional school performance, he was advised by his teacher to drop the subject. That led him to leave school and work as a printer's assistant.

By age 18, Mansfield had developed a keen interest in rocketry and applied for a job with the Rocket Propulsion Department of the UK Ministry of Supply. He stayed in this job for 18 months until he was called up for two years of mandatory army service. After serving two years in the British Army, Mansfield returned to the rocket propulsion department in 1954.

He also began to take night classes to gain a university placement. In 1956, he enrolled in a bachelor of science program in physics at Queen Mary College, University of London. He graduated in 1959 and stayed at Queen Mary College to study for his PhD. There, he worked in the magnetic resonance research group. In 1962, Mansfield graduated with a PhD in physics. In 1964, he became a lecturer at the University of Nottingham.

Mansfield followed Damadian's and Lauterbur's work closely but considered the length of time required for the MRI machines to produce an image a significant problem. In 1977, he created a new technique that allowed MRI scans to take seconds rather than hours and produced clearer images.

After failing to attract funding for his research, Damadian decided to set up his own company, called the Fonar Corporation, in 1978. Fonar aimed to produce and sell MRI machines, adopting techniques developed by Lauterbur and Mansfield. In 1980, the company sold the first MRI machine. Soon, Damadian's machines were in hospitals and laboratories all over the world. In the 1980s, he collaborated with fellow hero of progress Wilson Greatbatch (see chapter 37), who invented the implantable pacemaker, to create an MRI-compatible pacemaker.

In 1988, President Ronald Reagan awarded the National Medal of Technology to Damadian and Lauterbur for "their independent contributions in conceiving and developing the application of magnetic resonance technology to medical uses, including whole-body scanning and diagnostic imaging." Less than a year later, Damadian was inducted into the National Inventors Hall of Fame. In 2007, Lauterbur was honored in the same way.

In 2003, controversy arose when the Nobel Prize in Physiology and Medicine was presented to Lauterbur and Mansfield.

Despite Nobel rules allowing awards to be shared by up to three people, Damadian was not given the prize. Some have suspected that Damadian's creationist views, the fact that he was a physician and not an academic scientist, or his supposedly abrasive personality may have been a factor that contributed to his not being awarded the prize. In response to Nobel's announcement, Damadian took the unusual step of protesting the decision and took out several full-page advertisements in prominent newspapers all over the world to argue that he was deserving of the prize. Various MRI scientists have supported Damadian's claim to the Nobel Prize, but many other scientists criticized his response to the decision, calling it unprofessional.

Throughout his life, Lauterbur received dozens of awards and several honorary degrees. In 2007, he died, at the age of 77, from kidney disease at his home in Illinois. Mansfield also received a plethora of awards, including a knighthood in 1993 and the Lifetime Achievement Award, which was presented to him by the Prime Minister of the United Kingdom, Gordon Brown, in 2009. In 2017, Mansfield died, age 83, in Nottingham, England. Today, Damadian remains chairman of the board of Fonar and still lives in New York.

Thanks to the work of Damadian, Lauterbur, and Mansfield, the field of diagnostic medicine was changed forever. Without Damadian, it wouldn't be known that serious diseases could be detected by magnetic resonance. Without Lauterbur, there wouldn't be a way to clearly visualize the machine's results. And without Mansfield, MRI machines would take hours, rather than seconds, to scan patients. MRI scanners are among the most reliable diagnostic tools in all of medicine. Thanks to these machines, millions of lives have been extended and saved.

47

Yuan Longping

Inventor of hybrid rice

Yuan Longping was a Chinese agronomist who is known as the father of hybrid rice. In the early 1970s, he developed the first variant of high-yield hybrid rice. Yuan's discoveries, coupled with breakthroughs in wheat hybridization in the 1950s and 1960s by fellow hero of progress Norman Borlaug (see chapter 39), helped usher in the Green Revolution, which reduced the likelihood of famine in most of the world. Rice is the main staple food for approximately half the world's population. By increasing the plant's yield, Yuan's work helped save millions of lives. By the late 1990s, the increased yield brought about by Yuan's hybrid rice fed an additional 100 million Chinese people each year. Today, varieties of Yuan's rice are grown in more than 60 countries around the world.

Yuan was born in China on September 7, 1930, in Beiping, as Beijing was called at the time (the city was then commonly referred to in English as Peking). The Chinese Civil War (1927–1949), the Second Sino-Japanese War (1937–1945), and the associated economic turmoil forced Yuan's family to move repeatedly around southern China during his early life. Despite the disruptive upbringing, Yuan and his five siblings received a good education as both of their parents were teachers.

In 1949, Yuan finished high school and began studying at the Southwest Agricultural College (now Southwest University) on the outskirts of Chongqing in Sichuan province. Yuan's enrollment in college coincided with the victory of the Communist Party of China in the civil war and the party's consolidation of power across the country. For Yuan, who chose to major in agronomy with a focus on crop genetics, the new Chinese leadership presented a problem.

In the early 1950s, there were two primary theories of heredity. The first theory, based on the work of Soviet agronomists such as Ivan Vladimirovich Michurin and Trofim Lysenko, rejected modern genetics and proposed that organisms change over the course of their lives to adapt to altering environmental conditions. Champions of this idea claimed that by modifying a crop's environmental conditions—such as temperature, exposure to ultraviolet rays, and soil conditions—the plant's offspring would inherit these changes and eventually produce higher yields. Tragically, Lysenko's grain varieties were planted widely throughout the Soviet Union, resulting in massive crop failures and widespread hunger.

The other theory of heredity came from Western scientists such as the Austrian biologist and Augustinian friar Gregor

Mendel—often called the father of genetics—whose pea plant experiments established many rules of heredity still acknowledged today, and the American evolutionary biologist Thomas Hunt Morgan, who received the 1933 Nobel Prize in Physiology or Medicine for his discovery of the role chromosomes play in heredity. Mendel and Morgan believed that understanding genes was essential to understanding heredity. They proposed that while a set of genes is specific to each species, variations between individuals of the same species are also heritable—passed down from parents to their offspring—and occur because of the form each gene takes.

At that time, the Chinese government looked to the Soviet Union for nearly all scientific and technological insight, and the Soviet theory of heredity was accepted as truth in China. Anyone championing alternative scientific ideas could be charged with spreading misinformation and branded as a counterrevolutionary, which could lead to imprisonment or even death.

At college, Yuan was officially taught the theory of heredity championed by the Soviets. However, outside class, one of his professors, Guan Xianghuan, who rejected Soviet dogma, privately taught Yuan Western scientific theories. Guan encouraged Yuan to carry out experiments to test both Soviet and Western ideas. Although exposure to the Western ideas of heredity served Yuan well for his future career, a few years after mentoring Yuan, Guan was labeled an enemy of the Communist Party for his Western views. After years of harassment from the government, Guan took his own life in 1966.

In 1953, Yuan graduated from college and was assigned to teach crop cultivation, breeding, genetics, and Russian at Anjiang

Agricultural School, a small college in rural Hunan. During this time, Yuan conducted experiments to modify crops according to the Soviet theories of heredity. However, when these proved unsuccessful, he secretly read Western scientific magazines on crop science and changed his experiments to test Western methods.

In the late 1950s, Yuan's work on crop genetics and breeding became far more urgent. Between 1958 and 1960, Communist Party Chairman Mao Zedong's so-called Great Leap Forward directed the government to collectivize agriculture and plunged China into the worst famine of modern times. Nationwide, tens of millions died from starvation. Yuan was living in the Hunan countryside and saw the impact of the famine firsthand.

On roadsides, Yuan saw the bodies of several people who had died from starvation. "There was nothing in the field because hungry people took away all the edible things they could find," Yuan recalled later. "Famished, you would eat whatever there was to eat, even grassroots or tree bark. . . . I became even more determined to solve the problem of how to increase food production so that ordinary people would not starve."

In 1960, after two years of studying the sweet potato, Yuan switched to researching how to modify rice to create higher-yielding variants. According to Yuan, he switched to rice because it was the staple food of China, and the government was more likely to support and fund rice research.

In the West, the hybridization of wheat and maize led to tremendous food production breakthroughs and helped feed millions of people. Crop hybridization involves crossbreeding genetically dissimilar crops to produce offspring. A hybrid plant is usually more productive and can exhibit greater biomass, growth, or fertility than either parent plant, a phenomenon known as heterosis. Because rice is a self-pollinating plant, most scientists believed rice hybridization and heterosis were impossible.

In 1961, Yuan searched rice fields for months and eventually found what he considered an "outstanding" rice plant with large panicles and full grains. He meticulously collected more than 1,000 seeds of this rice and planted them the following year. To Yuan's surprise, the good traits of the parent crop were not passed down to the next generation.

After careful analysis, Yuan concluded that this rice was a natural hybrid. This discovery meant that, contrary to conventional thought, rice could be hybridized.

Despite this important discovery, one of the biggest problems facing Yuan was that hybridization requires different male and female plants as parents. As rice is self-pollinating, if the male parts of the rice plant were removed, crossbreeding and hybridization could occur when the remaining female parts accepted foreign pollen from other rice varieties. Unfortunately for Yuan, the time-consuming and delicate process of removing the male parts of the rice plant made that process impractical on a large scale. This led Yuan to hypothesize that if he and his team could find a strain of naturally mutated male-sterile rice (rice with female parts only), those plants could be used to hybridize new rice varieties.

In 1964, Yuan and a student spent the summer searching rice fields for these elusive naturally mutated male-sterile rice plants. In a 1966 article in the *Chinese Science Bulletin*, Yuan reported that he found six individual rice crops that had the potential for hybridization.

Unbeknownst to Yuan, this 1966 publication likely saved his life. At the time, the Cultural Revolution was in full force. Posters denouncing Yuan as a counterrevolutionary began appearing across his university, and local officials made plans to imprison him. The state even reserved a spot for him in the "cowshed," a place in the local prison for dissenting intellectuals. Fortunately, upon reading his publication about naturally mutated male-sterile rice, the director of the national science and technology commission and other provincial and national leaders sent a letter of support to the college in favor of Yuan's work. After that letter, Yuan was allowed to continue his work and was even provided greater financial support.

Despite having the six naturally mutated male-sterile rice plants, which he wrote about in his 1966 article, Yuan discovered that when these female plants were hybridized with pollen from other rice strains, their male-sterile traits were not passed down to their offspring. If Yuan couldn't find a way to ensure that the offspring of the hybridized rice passed on female parts only (that is, it was male-sterile), widespread hybridization would be extremely impractical. So Yuan and his team began searching for wild varieties of male-sterile rice, which Yuan thought may exhibit more promising genetic material.

In 1970, beside a railway line in Hainan Island, at long last, Yuan discovered a male-sterile wild rice plant that scientists refer to as "wild abortive." Soon after this discovery, he published a paper that outlined how the genetic material from the male-sterile wild rice could potentially be transferred into commercial rice strains. Yuan hypothesized that if that were done, the plant's offspring would still be male-sterile, and the world's heavily inbred commercial rice strains could be hybridized to produce greater yields.

In 1973, Yuan began harvesting the offspring of the wild abortive rice. To his delight, the offspring comprised tens of thousands of male-sterile rice plants. By the late 1970s, Yuan's hybrid rice was producing yields 20 to 30 percent higher than traditional commercial varieties.

This discovery of high-yield hybrid rice helped to alleviate food insecurity not only in China but around the world. In doing so, Yuan's rice saved millions of lives.

Throughout the 1980s, Yuan donated his key rice varieties to various domestic and international agronomists and organizations for no profit. He and his team trained farmers and introduced his hybrid strains in more than 80 countries.

By 1991, the United Nations found that 20 percent of the world's rice output came from the 10 percent of the world's rice fields that grew Yuan's hybrid rice. In 1999, scientist Dennis Normile found that in China alone, the production increases brought about by hybrid rice fed an additional 100 million people each year. Today, one-fifth of all rice grown globally originates from Yuan's hybrids.

Later in life, Yuan became something of a national celebrity in China and was praised extensively by the ruling Communist Party. However, quite unusually for someone of his prominence, he never engaged in politics or joined the Communist Party. Yuan was awarded dozens of national and international awards. In 2000, he was awarded the UNESCO Prize for Science, and in 2004, he was awarded the World Food Prize. Four asteroids, a minor planet, and a college in China are named after him.

Yuan continued his work to make his hybrid varieties even more productive. These advancements included crossbreeding rice with maize to be more nutritious and enriching rice with vitamin A to help improve people's vision. As recently as 2018, at the age of 87, Yuan and his team created a hybrid rice variety that could grow in salt-rich soil, which helps farmers living in coastal areas. Yuan died on May 22, 2021, in Changsha, Hunan. In the days following his death, thousands of mourners lay flowers and bowls of boiled rice (a sign of respect for the deceased in some parts of Asia) outside the funeral home.

By developing the first variant of high-yield hybrid rice, Yuan's work has improved the world's food stability. He persevered and proved the naysayers wrong, despite fierce political resistance early in his career and critics who believed that rice hybridization was impossible. As a result, he has saved millions of lives, and millions of people eat Yuan's hybrid rice every day.

48

Tu Youyou

Malaria treatment pioneer

Tu Youyou is the scientist who discovered artemisinin, the core compound used to create extremely effective malaria-fighting drugs. Her work is considered a breakthrough in 20th-century tropical medicine. Since artemisinin's discovery, it has been used to save tens of millions of lives worldwide.

Tu Youyou was born on December 30, 1930, in Ningbo, a city on the east coast of China. As a child, she was fortunate enough to attend some of the top private schools in the region. At the age of 15, Tu contracted tuberculosis and had to take a two-year break from her studies. Fortunately for humanity, her illness inspired her to go into medicine, a profession where she could try to find cures for diseases like the one that had afflicted her.

Tu returned to school in 1948. In 1951, she began studying at Peking University Medical School (now called Peking University Health Science Center). Four years later, Tu was assigned to continue her research at the newly established Academy of Traditional Chinese Medicine. As her undergraduate degree focused primarily on Western medicine, she took a full-time course in traditional Chinese medicine between 1959 and 1962.

Tu was doing her research during China's Cultural Revolution in the 1960s and 1970s, a time when scientists and intellectuals were often vilified by the Chinese government, and many were imprisoned, executed, or sent to "reeducation camps." The government also shut down hundreds of research programs. For a brief period, Tu's husband, an engineer, was taken by the government to a reeducation camp.

Before Tu's discovery, malaria was effectively treated with the drugs chloroquine and quinoline. However, in the late 1960s, new strains of malaria evolved and became resistant to the existing drug treatments. The disease quickly spread as the global medical community struggled to respond to the new strains, resulting in the deaths of millions of people.

Malaria's resurgence was particularly catastrophic in Southeast Asia, where it afflicted the forces engaged in the Vietnam War. As Tu noted in her 2015 Nobel Lecture, "casualties in the U.S. military force [*sic*] caused by medical disability due to the full seasonal prevalence of malaria reached four to five times higher than casualties from actual direct combat in 1964." Thus, tackling malaria quickly became a top medical priority for both parties in the conflict.

In 1967, Ho Chi Minh, the leader of North Vietnam, pleaded with Zhou Enlai, the Chinese premier, for a new malaria treatment for his soldiers. The early part of Tu's career had been centered on finding a cure for schistosomiasis, a disease caused by a parasitic flatworm. In 1967, Tu was approached to join a top-secret government drug discovery program named Project 523, which would focus on developing a cure for malaria.

In 1969, Tu was appointed the head of her research group and was sent to the Hainan region to study patients who had been infected with malaria. She was forced to leave behind her one- and four-year-old daughters. It would be three years before she saw them again. Looking back at that time, Tu said, "the work was the top priority, so I was certainly willing to sacrifice my personal life."

After scientists from around the world had unsuccessfully screened more than 240,000 compounds for their effectiveness against malaria, Tu thought it could be useful to try Chinese herbs. By 1971, Tu and her team had tested over 2,000 traditional Chinese recipes. After searching through dozens of history books, her team found a concoction from a book from the year 400 CE titled *Emergency Prescriptions Kept Up One's Sleeve* that used the ingredient sweet wormwood to treat intermittent fevers, a hallmark symptom of malaria.

At first, sweet wormwood proved ineffective against malaria. However, Tu found inspiration from another traditional Chinese medical book, *The Handbook of Prescriptions for Emergency Treatments*, written in 340 CE by Ge Hong. Tu realized that instead of boiling the sweet wormwood to extract its antimalarial properties, she should instead attempt a low-temperature extraction. Early tests on mice and monkeys proved to be 100 percent successful.

In 1972, Tu prevailed in extracting the pure antimalarial substance from sweet wormwood and named it *qinghaosu,* or artemisinin, as it is commonly known in the West. Tu insisted that she should be the first human test subject. Then, she tested her discovery on 21 patients. Artemisinin proved to be fully effective in treating malaria patients. Tu published her findings anonymously in 1977, and artemisinin was soon used in anti-malaria drugs globally.

In 1980, Tu was promoted to researcher, equivalent to the academic rank of full professor in the West. In 1981, she presented her findings on artemisinin at a meeting of the World Health Organization. She continues to work as the chief scientist at the Academy of Traditional Chinese Medicine, where she has worked since 1955.

Tu has been decorated with numerous awards. Most notably, she was awarded the Lasker Award in clinical medicine in 2011. In 2015, she was one of three people to win the Nobel Prize in Physiology or Medicine. Tu was the first Chinese Nobel laureate in physiology or medicine and the first Chinese woman to receive a Nobel Prize in any category.

Because of the millions of lives artemisinin has saved, the Lasker Foundation described Tu's discovery as "arguably the most important pharmaceutical intervention in the last half [of the 20th] century."

49

Françoise Barré-Sinoussi and Luc Montagnier

Trailblazing AIDS researchers

Luc Montagnier and Françoise Barré-Sinoussi are French scientists who discovered that HIV is the cause of AIDS. The pair's discovery has led to the development of medical treatments that slow the progression of HIV and decrease the risk of the virus's transmission.

Luc Antoine Montagnier was born on August 18, 1932, in Chabris, a small commune in Centre-Val de Loire (Loire Valley), France. As a teenager, after witnessing his grandfather suffer and eventually die from cancer, he decided to become a medical researcher and focus on cancer. In high school, Montagnier was a bright student, and his passion for science led him to set up a small chemistry laboratory in the cellar of his parents' house. Looking back on his homemade lab, Montagnier declared, "[T]here, I enthusiastically produced hydrogen gas, sweet-smelling aldehydes, and nitro compounds."

After finishing his preparatory education, Montagnier began study at the University of Poitiers. Initially, he wanted to concentrate on human biology; however, there was no such specialty at Poitiers, so he compromised by studying for a science degree. To gain experience in medical research, he spent his mornings working at a local hospital.

In 1953, Montagnier graduated from Poitiers and soon began studying medicine at the University of Paris. After graduating with a degree in medicine in 1960, he devoted the early part of his career to cancer research. He spent the next 12 years working at various medical research institutes in Paris, London, and Glasgow.

In 1972, Montagnier began working at the Pasteur Institute, a Parisian research center studying biology, diseases, and vaccines. There, Montagnier set up the institute's Department of Virology, a research unit dedicated to detecting viruses involved in human cancers. During his time at the Pasteur Institute, Montagnier and his colleague Françoise Barré-Sinoussi made their groundbreaking discovery about HIV.

Françoise Barré-Sinoussi was born on July 30, 1947, in Paris. She showed an interest in science from an early age and decided to continue her passion for knowledge at the University of Paris. Like Montagnier, Barré-Sinoussi initially wanted to study medicine. However, as she came from a humble background, she decided to be pragmatic and study natural science. Natural science was a shorter course than a medical degree, which saved her family money in tuition and boarding fees.

After studying at the University of Paris for a couple of years, Barré-Sinoussi began working part-time at the Pasteur Institute. She soon began working full-time at the institute and attended the university only to take her exams. Barré-Sinoussi received a PhD in 1975. After a brief internship at the National Institutes of Health in Bethesda, Maryland, she began working with Montagnier at the Pasteur Institute, researching a group of viruses known as retroviruses.

During the 1980s AIDS epidemic, scientists were perplexed as to what had caused the outbreak of the disease. In 1982, Willy Rozenbaum, a clinician at Bichat-Claude Bernard Hospital in Paris, asked Montagnier for assistance in identifying the cause of this mysterious new disease. Montagnier and Barré-Sinoussi began working immediately, and within a year, they made the groundbreaking discovery that HIV caused AIDS. Montagnier and Barré-Sinoussi published their findings in an article in *Science* on May 20, 1983.

Their discovery led to many medical breakthroughs that have helped in the fight against AIDS, including numerous HIV testing and diagnosis technologies and lifesaving antiretroviral therapies.

In 1985, Montagnier became a professor in the Department of AIDS at the Pasteur Institute. In 1993, he established the World Foundation for AIDS Research and Prevention. Since then, Montagnier has worked at various universities, including Queens College in New York and, more recently, Shanghai Jiao Tong University.

In 1988, Barré-Sinoussi took charge of her own laboratory at the Pasteur Institute and began intensive research on creating an HIV vaccine. Since 2006, Barré-Sinoussi has worked as the president and then cochair of the International AIDS Society. Although a vaccine has yet to be discovered, her team continues to research various mechanisms to protect people against HIV infection.

In 2008, Barré-Sinoussi and Montagnier were awarded the Nobel Prize in Physiology or Medicine for their work in discovering HIV as the cause of AIDS. They shared the prize with Harald zur Hausen, who found that human papillomavirus, or HPV, can cause cervical cancer.

Montagnier has received dozens of awards, including the Lasker Award, *Commandeur de l'Order National du Mérite* (a French order of merit), and numerous professional honors from countries around the world. In 2000, he was portrayed on a stamp issued by Bhutan. Similarly, Barré-Sinoussi has received many awards and honorary doctorates, including the Prize of the French Academy of Sciences and the Grand Officier de la Légion d'Honneur, France's highest order of merit.

Thanks to the discovery of HIV and the development of treatments, humanity is now winning the war on AIDS. At the peak of the HIV pandemic in the mid-2000s, some 1.9 million people died of AIDS each year, but fewer than one million people died from the disease in 2017. There are also fewer new infections. In the mid-1990s, there were 3.4 million new HIV infections each year, but in 2017, there were only 1.8 million new HIV infections—a decline of 47 percent.

Without the contributions of Montagnier and Barré-Sinoussi, the fight against AIDS would not be as advanced or successful as it is today, and millions more people would be dying from the virus each year.

50

David Nalin

Oral rehydration therapy formula developer

David Nalin is a scientist who developed the precise formula for oral rehydration therapy (ORT). He led the team that carried out the first successful trials of ORT and helped to spread its use around the world. ORT is used to rehydrate patients suffering from illnesses that cause severe dehydration. Our World in Data, a project of the University of Oxford, has estimated that since its creation in 1968, ORT has saved more than 70 million lives.

Before ORT was established, the only effective way to rehydrate a patient suffering from serious dehydrating illnesses, such as cholera, was to provide fluids intravenously. Intravenous (IV) therapy is an expensive and frequently inaccessible treatment, as it requires modern medical facilities that are often unavailable in developing nations. If left untreated, cholera can kill a healthy person in just a few hours. That is where David Nalin enters the story.

David Nalin was born on April 22, 1941, in New York City. He attended the Bronx High School of Science, and in 1957 he enrolled in a bachelor of arts zoology program at Cornell University. After graduating from Cornell in 1961, he secured a place at Albany Medical College and four years later earned his medical doctorate.

After spending a year as a medical intern at Montefiore Hospital in New York, in 1967 Nalin moved to Dhaka, the capital of East Pakistan (now Bangladesh), where he worked on cholera for the Southeast Asia Treaty Organization, or SEATO. The following year, a large cholera epidemic broke out in Dhaka. Nalin quickly realized that the existing IV therapy that was used to treat cholera was too costly and slow to administer. Working from a small missionary hospital in the jungle, Nalin led a small team of researchers to develop and complete the first successful trial for ORT.

ORT mixes salts and sugars with water to replace the minerals lost by the patient because of diarrhea or vomiting. The salts are needed for organ function, and the sugars help the salts to be absorbed in the intestines. Nalin acknowledged that ORT is a very simple solution, but the nuance in his discovery comes from the very specific ratios of water, salt, and sugar needed for the treatment to be effective. ORT packets are very cheap and cost just between three and four U.S. cents to produce. The solution doesn't need to be administered by medical professionals, and it uses between 70 and 80 percent less fluid than is needed with IV therapy.

In 1969, Nalin began working as a consultant for the World Health Organization. He helped to successfully establish ORT programs to combat diarrheal diseases around the world, including in Costa Rica, Jamaica, Jordan, and Pakistan. In 1971, during the Bangladesh War of Independence, ORT helped reduce the cholera mortality rate from 30 percent to just 3.6 percent.

In 1978, the prestigious British medical journal *The Lancet* called ORT "potentially the most important medical advance of this century." Similarly, in 1987, the United Nations Children's Fund (UNICEF) noted that "no other single medical breakthrough of the 20th century has had the potential to prevent so many deaths over such a short period of time and at so little cost" as ORT.

From 1983 until his retirement in 2002, Nalin held several director positions at Merck Research Laboratories. In 2002, he received the first Pollin Prize in Pediatric Research, awarded by New York-Presbyterian Hospital. In 2007, he received the Prince Mahidol Award for outstanding achievements in medicine and public health from the King of Thailand. Since his retirement, Nalin has continued to work as a consultant in vaccinology. He is currently a professor emeritus at Albany Medical College.

In addition to the 70 million lives that ORT has already saved, Nalin's formulation continues to save thousands more lives every day.

51

Alfred Sommer

Pioneering vitamin researcher

Alfred Sommer is an American scientist who discovered that vitamin A deficiency dramatically contributes to increased childhood mortality and morbidity—a discovery that led to significant advances in child nutrition and health. Sommer found that providing children, especially those living in the developing world, with vitamin A tablets twice a year could drastically reduce the rate of childhood death and blindness. Sommer's finding and treatment have saved more than 10 million children from an early death and continue to save hundreds of thousands more people every year.

Alfred Sommer was born in New York City on October 2, 1942. He attended Union College in Schenectady, New York, where he graduated summa cum laude with a bachelor of science in biology and a minor in history in 1963. After finishing his bachelor's degree, Sommer attended Harvard Medical School, where he obtained an MD in 1967.

After graduating from Harvard, Sommer worked as a medical intern at Harvard's Beth Israel Deaconess Medical Center. In 1969, he moved with his family to Dhaka, East Pakistan (now Bangladesh), to research cholera for the Centers for Disease Control and Prevention.

In 1972, Sommer returned to the United States to continue his education. In 1973, he completed a master of health sciences in epidemiology at the Johns Hopkins School of Hygiene and Public Health and began to work as a fellow in ophthalmology at the Johns Hopkins Wilmer Eye Institute, a clinic specializing in the diagnosis and management of complex eye diseases. Here, he began to research the causes and effects of vitamin A deficiency.

In 1976, Sommer moved to Indonesia to study the problem more closely. He became a visiting professor of ophthalmology at Padjadaran University, where he soon discovered that vitamin A deficiency was the leading cause of preventable blindness in children.

He also discovered that vitamin A deficiency dramatically increased childhood mortality rates by reducing patients' resistance to infectious diseases. Soon after his discovery of the debilitating effects of vitamin A deficiency, Sommer found that he could treat the condition cheaply and effectively by providing patients with a twice-yearly, oral high-dose vitamin A supplement.

To convince the scientific community of the importance of vitamin A deficiency, Sommer and his colleagues ran numerous large-scale randomized trials. He organized an international conference on the topic at the Rockefeller Foundation's Bellagio Center in Italy. At the event, scientists concluded that the use of twice-yearly doses of vitamin A could reduce the mortality rate of vitamin A-deficient children by 34 percent and save one million children from blindness every year.

Importantly, from a trial conducted in Nepal, Sommer also found that giving vitamin A supplements to women of child-bearing age reduced maternal mortality by 45 percent.

Because of how cheap and easy it is to provide semi-annual vitamin A supplements to those in need, the World Bank's *World Development Report* declared Sommer's vitamin A supplementation one of the most cost-effective of all health interventions.

In 1980, Sommer returned to Johns Hopkins Wilmer Eye Institute as the founding director of the Dana Center for Preventive Ophthalmology. Today, he continues to focus on clinical epidemiology, blindness prevention, and child survival. Sommer is currently dean emeritus and professor of epidemiology and international health at Johns Hopkins Bloomberg School of Public Health and Hygiene.

Throughout his career, Sommer has been decorated with numerous awards, including the Pollin Prize in Pediatric Research and the Prince Mahidol Award for outstanding achievements in medicine and public health from the King of Thailand. He is also an elected member of both the National Academy of Sciences and the National Academy of Medicine.

Thanks to the work of Alfred Sommer, tens of millions of people have been saved from blindness, and more than 10 million lives have been saved to date.

52

Bill Gates

Computer entrepreneur and philanthropist

Bill Gates is an American software developer, businessman, and philanthropist. Thanks to Microsoft, the company he cofounded, personal computers went from being used almost exclusively by computer hobbyists to being a staple in millions of homes and offices around the world. By helping to make computers accessible to the masses, Gates's work dramatically changed the world we live in and made many once-complicated tasks much simpler. Moreover, global productivity increases enabled by his company's software have likely added trillions of dollars to the world economy.

Gates has also created several charitable organizations. He and his then-wife established the Bill and Melinda Gates

Foundation, the world's largest philanthropic organization, in 1994. By providing education, vaccinations, and investment in infrastructure; combating diseases; and building improved sanitation facilities, the Gates Foundation's philanthropic endeavors have improved the lives of hundreds of millions of the world's poorest people and helped to save tens of millions of lives.

William Henry Gates was born on October 28, 1955, in Seattle, Washington. His father was a well-known lawyer, and his mother served on the board of several large companies. Gates grew up in a household he would later describe as "well-off," and his two sisters have described him as a happy child. At an early age, he showed signs of competitiveness and intelligence. He excelled at playing the board games Risk and Monopoly and would often spend hours a day reading.

At 13 years old, Gates began studying at Lakeside School, an exclusive private school in Seattle. He performed exceptionally in nearly all his subjects but had aptitudes for math and science. He later noted that through the eighth grade, he enjoyed the fact that he was able to do reasonably well in school without much effort.

Soon after Gates arrived at Lakeside, the school obtained a teletype terminal (an electromechanical teleprinter). A local computer company also offered Lakeside students some time on its computers. Gates jumped at the opportunity to use these computers and quickly became fascinated with computer software. He spent much of his free time working on the teletype terminal and was even excused from other lessons to pursue this interest. During this time, Gates wrote his first computer program: a game of tic-tac-toe.

Gates forged several friendships in the school's computer room, notably with Paul Allen, eventually Microsoft's other cofounder; Ric Weiland, Microsoft's first employee; and Kent Evans, Gates's best friend and first business partner. One summer, Gates and his friends were banned from the local computer shop after they were caught exploiting bugs in the system to gain free computer time. After the ban, the four students formed the Lakeside Programmers Club to make money. The club volunteered to help the computer store fix the bugs in its machines' software.

In 1971, Gates and Evans began to automate Lakeside's class-scheduling system. Unfortunately, after the end of the school year, Evans died in a mountain climbing accident. Gates, deeply saddened by Evans's passing, turned to Allen to help finish the project.

Although Allen was two years older than Gates, and despite their not always seeing eye to eye, the teenagers quickly became friends and bonded over their passion for software. When Gates was 17, he and Allen formed their first business venture, Traf-O-Data, a computer program that helped monitor traffic patterns in Seattle. The pair earned $20,000 for their work.

In 1973, Gates graduated from Lakeside with an SAT score of 1590 out of 1600. In the fall, he enrolled at Harvard University. Although Gates's major was pre-law, he spent most of his time at Harvard studying mathematics and graduate-level computer science.

In 1975, Gates read about the new Altair 8800 microcomputer kit in the magazine *Popular Electronics*. Gates and Allen decided to write to the company that had created the Altair, Micro Instrumentation and Telemetry Systems (MITS), to gauge the company's interest in having someone build a software program that could run on the computer. To do that, Gates decided to call Ed Roberts, the president of MITS, from his dorm phone.

Gates told Roberts that he had already created the software and attempted to sell MITS the program. In reality, neither Gates nor Allen had an Altair machine, and the pair hadn't written a single line of code for the computer. Roberts asked for a demonstration, and Gates and Allen spent the next month working around the clock in Harvard's computer lab writing the code. "From that moment," Gates later remembered, "I worked day and night on this little extra credit project that marked the end of my college education and the beginning of a remarkable journey with Microsoft."

In the spring of 1975, Allen and Gates traveled to MITS's offices in New Mexico to test run the code. They had never tested it before, but the demonstration was a success, and both Allen and Gates were hired by MITS. Gates never returned to study at Harvard.

While working at MITS, Gates and Allen also formed a partnership, which they named "Micro-Soft"—a combination of the words "microcomputer" and "software." The company focused on developing programming language software for a variety of different systems. By 1976, Gates and Allen left MITS, dropped the hyphen in their company's name, registered the name Microsoft, and opened their first office in Albuquerque, New Mexico. The same year, they hired their old high school friend Ric Weiland as Microsoft's first employee.

By 1979, Microsoft was grossing approximately $2.5 million a year. In 1980, Microsoft struck a deal with IBM to provide the basic operating system that would run on IBM's new computers. Although the contract with IBM earned Microsoft only a small fee, the prestige of doing business with one of the world's largest corporations helped transform Microsoft into one of the world's leading software companies.

In 1981, Microsoft, formerly a partnership between Allen and Gates, was reorganized as a privately held corporation. At age 23, Gates was made CEO and chairman of the board, and the offices moved to Bellevue, Washington. For the first five years of Microsoft's existence, Gates personally reviewed and often rewrote every line of code the company's programmers created. By 1983, about 30 percent of the world's computers ran on Microsoft software, and the company established subsidiaries in England, Japan, and France. In the same year, Allen left Microsoft.

In November 1985, Microsoft released the first retail version of Microsoft Windows. The program sold well, and in 1986, when Microsoft went public, Gates became the world's youngest billionaire. Since then, he has continually been one of the world's richest people.

In 1989, Gates released Microsoft Office, which included the early versions of applications such as Microsoft Word and Excel. In 1995, Microsoft released the operating system Windows 95. The software marked an enormous leap forward in terms of both graphics and, more important, the design of operating systems and sold rapidly.

Thanks to the Windows operating system, computers were no longer too complicated for the everyday person to use.

Internet Explorer was released a few weeks after Windows 95. For the first time, millions of people began to use the internet. A personal computer revolution ensued in the years that followed, as the price of computers dropped and computer ownership skyrocketed. Microsoft remains one of the largest corporations in the world.

In 2000, Gates stepped down as CEO of Microsoft. Since then, he has focused most of his efforts on the Bill and Melinda Gates Foundation, which he set up with his then-wife in 1994. The Foundation is involved in many fields, including tackling poverty in Washington state, increasing access to basic sanitation facilities in the developing world, educating women, reducing the number of HIV infections and extending the lives of those infected, tackling malaria, and working to completely eradicate polio. Since 1994, Bill and Melinda Gates have donated more than $50 billion to a variety of causes.

In 2000, the Bill and Melinda Gates Foundation created Gavi, the Vaccine Alliance, which has helped vaccinate more than 760 million children and has prevented more than 13 million deaths. All in all, the Bill and Melinda Gates Foundation has improved the lives of hundreds of millions of the world's poorest people.

During a TED talk in 2015, Gates famously warned that the world was not prepared for the next pandemic. Over the past few years, he has spent millions of dollars on novel-virus preparedness. In addition, during the COVID-19 pandemic, Gates invested hundreds of millions of dollars in efforts to curb the outbreak.

Gates and his close friend Warren Buffett also created a campaign called the Giving Pledge, which encourages the extremely wealthy to donate the majority of their wealth to charitable causes. So far, 204 people have signed the campaign, pledging a total of $1.2 trillion to charity.

Throughout his career, Gates has received dozens of awards and honors. *Time* magazine named Gates one of the most influential people of the 20th century. In 2005, he was given an honorary knighthood by the United Kingdom's Queen Elizabeth II. In 2016, he and Melinda Gates were awarded the Presidential Medal of Freedom. Gates also holds many honorary doctorates from universities across the world, including Harvard, from which he dropped out.

Thanks to Bill Gates, computers went from being used almost exclusively by hobbyists, who spent hours learning the complex languages necessary to operate a personal computer, to a vital, easy-to-use product used by billions of people. And through his philanthropy, Gates has saved tens of millions of lives and helped to improve hundreds of millions more.

Suggested Further Reading

Ronald Bailey and Marian L. Tupy, *Ten Global Trends Every Smart Person Should Know: And Many Others You Will Find Interesting* (Washington: Cato Institute, 2020).

Angus Deaton, *The Great Escape: Health, Wealth, and the Origins of Inequality* (Princeton, NJ: Princeton University Press, 2013).

Chelsea Follett, *Centers of Progress: 40 Cities That Changed the World* (Washington: Cato Institute, 2023).

Deirdre Nansen McCloskey and Art Carden, *Leave Me Alone and I'll Make You Rich: How the Bourgeois Deal Enriched the World* (Chicago: University of Chicago Press, 2020).

Johan Norberg, *Progress: Ten Reasons to Look Forward to the Future* (New York: OneWorld Publications, 2017).

Johan Norberg, *Open: Story of Human Progress* (London: Atlantic, 2020).

Steven Pinker, *The Better Angels of Our Nature: Why Violence Has Declined* (New York: Viking Press, 2011).

Steven Pinker, *Enlightenment Now: The Case for Reason, Science, Humanism, and Progress* (New York: Viking, 2018).

Matt Ridley, *The Rational Optimist* (New York: HarperCollins, 2010).

Matt Ridley, *The Evolution of Everything: How New Ideas Emerge* (New York: HarperCollins, 2015).

Matt Ridley, *How Innovation Works: And Why It Flourishes* (New York: HarperCollins, 2020).

Hans Rosling, Ola Rosling, and Anna Rosling Rönnlund, *Factfulness: Ten Reasons We're Wrong About the World—and Why Things Are Better Than You Think* (London: Sceptre, 2018).

Marian L. Tupy and Gale Pooley, *Superabundance: The Story of Population Growth, Innovation, and Human Flourishing on an Infinitely Bountiful Planet* (Washington: Cato Institute, 2022).

Discussion Questions for Book Clubs and Classrooms

1. What are the common characteristics of the societies that the heroes featured in this book usually came from?
2. How important is innovation to the modern world we know today?
3. What role does freedom of expression, and being able to publicly disagree with authority without persecution, play in promoting innovation and human progress?
4. What role did immigration play in providing our heroes opportunities to innovate?
5. Who do you think was the most significant hero documented in this book? Which hero improved the world the most?
6. What role do improvements on past innovations have in promoting human progress?
7. If you could have dinner with any one of these heroes, which one would you choose?
8. If you could include an additional chapter, which individual would you include? For what achievement or innovation?
9. If this book is rewritten in 100 years, which heroes or inventions do you think would be featured?
10. How, if at all, has this book changed your perspective?

Acknowledgments

Thank you to all members of the staff and donors to the Cato Institute, my former employer and this book's publisher. A special thanks goes to Marian Tupy for his many years of mentorship and friendship, for editing every chapter of this book, and for taking the chance to employ me in my first think tank job all those years ago. Thanks to Ian Vásquez for his belief in this project and continued support and to Chelsea Follett for her frequent suggestions of heroes to include and countless hours of meticulous editing. I am also grateful to Luis Ahumada Abrigo and Guillermina Sutter Schneider for formatting and designing this book, including its beautiful cover. I also thank Eleanor O'Connor for managing the creation of the book, and Ivan Osorio, Aaron Steelman, Sara Proehl, and Melanie Donohue for fantastic editorial suggestions.

I am forever grateful to Maria for her endless encouragement and for being a rock in my life and to James and Brunya for helping ensure that I never take life too seriously. A final thank-you goes to Mum and Dad, for their years of unrelenting support and for always pushing me to pursue my dreams, no matter how crazy they were. You are all heroes to me.

Index

About the Author

Alexander C. R. Hammond is the founder of the Initiative for African Trade and Prosperity, a free trade fellow at the Institute of Economic Affairs (IEA), and a senior fellow at African Liberty. His work focuses on the history of economic growth and innovation, African development, and global well-being. He was formerly a policy analyst at the IEA and a researcher in the Cato Institute's Center for Global Liberty and Prosperity.

Alexander's popular publications have been translated into more than a dozen languages and appeared in major newspapers across Europe, North America, and Africa. He received a BA with combined honors in history and politics from the University of Exeter and a master's degree by research in political economy from the University of Buckingham. Alexander currently resides in Switzerland.

About the Cato Institute and Human Progress

Founded in 1977, the Cato Institute is a public policy research foundation dedicated to broadening the parameters of policy debate to allow consideration of more options that are consistent with the principles of limited government, individual liberty, and peace. The Institute is named for *Cato's Letters*, libertarian pamphlets that were widely read in the American colonies in the early 18th century and played a major role in laying the philosophical foundation for the American Revolution.

The Cato Institute undertakes an extensive publications program on the complete spectrum of policy issues. Books, monographs, and shorter studies are commissioned to examine the federal budget, Social Security, regulation, military spending, international trade, and myriad other issues. Major policy conferences are held throughout the year.

HumanProgress.org, a project of the Cato Institute, was founded in 2013 to rectify the widely held misperceptions about the state of humanity by gathering empirical data from reliable sources that look at worldwide long-term trends. By compiling and presenting these comprehensive data in an accessible way, the editors aim to provide a useful resource for students, scholars, journalists, policymakers, and the general public.

To maintain its independence, the Cato Institute accepts no government funding. Contributions are received from foundations, corporations, and individuals, and other revenue is generated from the sale of publications. The Institute is a nonprofit, tax-exempt, educational foundation under Section 501(c)(3) of the Internal Revenue Code.